Lecture Notes in Mathematics

Edited by J.-M. Morel, F. Takens and B. Teissier

Editorial Policy for Multi-Author Publications: Summer Schools / Intensive Courses

1. Lecture Notes aim to report new developments in all areas of mathematics and their applications – quickly, informally and at a high level. Mathematical texts analysing new developments in modelling and numerical simulation are welcome. Manuscripts should be reasonably self-contained and rounded off. Thus they may, and often will, present not only results of the author but also related work by other people. They should provide sufficient motivation, examples and applications. There should also be an introduction making the text comprehensible to a wider audience. This clearly distinguishes Lecture Notes from journal articles or technical reports which normally are very concise. Articles intended for a journal but too long to be accepted by most journals, usually do not have this „lecture notes" character.

2. In general SUMMER SCHOOLS and other similar INTENSIVE COURSES are held to present mathematical topics that are close to the frontiers of recent research to an audience at the beginning or intermediate graduate level, who may want to continue with this area of work, for a thesis or later. This makes demands on the didactic aspects of the presentation. Because the subjects of such schools are advanced, there often exists no textbook, and so ideally, the publication resulting from such a school could be a first approximation to such a textbook.

   Usually several authors are involved in the writing, so it is not always simple to obtain a unified approach to the presentation.

   For prospective publication in LNM, the resulting manuscript should not be just a collection of course notes, each of which has been developed by an individual author with little or no co-ordination with the others, and with little or no common concept. The subject matter should dictate the structure of the book, and the authorship of each part or chapter should take secondary importance. Of course the choice of authors is crucial to the quality of the material at the school and in the book, and the intention here is not to belittle their impact, but simply to say that the book should be planned to be written by these authors jointly, and not just assembled as a result of what these authors happen to submit.

   This represents considerable preparatory work (as it is imperative to ensure that the authors know these criteria before they invest work on a manuscript), and also considerable editing work afterwards, to get the book into final shape. Still it is the form that holds the most promise of a successful book that will be used by its intended audience, rather than yet another volume of proceedings for the library shelf.

3. Manuscripts should be submitted (preferably in duplicate) either to Springer's mathematics editorial in Heidelberg, or to one of the series editors (with a copy to Springer). Volume editors are expected to arrange for the refereeing, to the usual scientific standards, of the individual contributions. If the resulting reports can be forwarded to us (series editors or Springer) this is very helpful. If no reports are forwarded or if other questions remain unclear in respect of homogeneity etc, the series editors may wish to consult external referees for an overall evaluation of the volume. A final decision to publish can be made only on the basis of the complete manuscript; however a preliminary decision can be based on a pre-final or incomplete manuscript. The strict minimum amount of material that will be considered should include a detailed outline describing the planned contents of each chapter.

   Volume editors and authors should be aware that incomplete or insufficiently close to final manuscripts almost always result in longer evaluation times. They should also be aware that parallel submission of their manuscript to another publisher while under consideration for LNM will in general lead to immediate rejection.

Continued on inside back-cover

Lecture Notes in Mathematics 1872

Editors:
J.-M. Morel, Cachan
F. Takens, Groningen
B. Teissier, Paris

Editor *Mathematical Biosciences Subseries:*
P.K. Maini, Oxford

Avner Friedman (Ed.)

Tutorials in
Mathematical Biosciences III

Cell Cycle, Proliferation, and Cancer

With Contributions by:

B. Aguda · M. Chaplain · A. Friedman
M. Kimmel · H.A. Levine · G. Lolas
A. Matzavinos · M. Nilsen-Hamilton · A. Swierniak

 Springer

mbi

Mathematical Biosciences Institute
at The Ohio State University

Editor

Avner Friedman

The Ohio State University
Mathematical Biosciences Institute
W . 18th Avenue 231
43210-1292 Ohio
U.S.A.

e-mail: afriedman@mbi.osu.edu

Library of Congress Control Number: 2005934038

Mathematics Subject Classification (2000): 34C60, 35G25, 35G30, 35M10, 35Q80, 35R30, 35R35, 49K15, 92C15, 92C17, 92C37, 92C50, 93C15

ISSN print edition: 0075-8434
ISSN electronic edition: 1617-9692
ISBN-10 3-540-29162-8 Springer Berlin Heidelberg New York
ISBN-13 978-3-540-29162-6 Springer Berlin Heidelberg New York

DOI 10.1007/11561606

Springer is a part of Springer Science+Business Media
springer.com
© Springer-Verlag Berlin Heidelberg 2006

Typesetting: by the authors and Techbooks using a Springer LATEX package

Cover design: *design & produktion* GmbH, Heidelberg

Printed on acid-free paper   SPIN: 11561606   41/Techbooks   5 4 3 2 1 0

Preface

This is the third volume in the series "Tutorial in Mathematical Biosciences". These lectures are based on material which was presented in tutorials or developed by visitors and postdoctoral fellows of the Mathematical Biosciences Institute (MBI), at The Ohio State University. The aim of this series is to introduce graduate students and researchers with just a little background in either mathematics or biology to mathematical modeling of biological processes. The first volume was devoted to mathematical neuroscience, which was the focus of the MBI program in 2002–2003. The second volume dealt with mathematical modeling of calcium dynamics and signal transduction, which was the focus of the MBI program in the winter of 2004. Documentation of these activities, including streaming videos of the workshops, can be found on the web site http://mbi.osu.edu.

The present volume is devoted to the topics of cell division cycle, tumor growth, and cancer chemotherapy. These topics were featured in three MBI workshops held during the fall of 2003. The first chapter gives an overview of the modeling of cell division cycle. This is a process of replicating the genetic material as well as other cellular components of the cell. The emphasis here is not on the biochemistry, but rather on the dynamics arising from the topology of the network of molecular interactions.

Chapters 2–4 deal with various aspects of tumor growth. At the early stage of tumor growth, the tumor cells receive nutrients (oxygen, glucose, etc.) from the blood, which circulates in the vasculature. However, as the tumor grows, the blood supply is unable to keep up with the demand: Indeed, cells which reside in the core of the tumor do not receive enough nutrients and they become necrotic. In order to enable its continued growth, the tumor secretes growth factors and, as a result, new blood vessels are formed and move into the tumor in a process called angiogenesis. The mathematical model and analysis of this process are described in Chap. 2. By the time a tumor has grown to a size that can be detected, there is a strong likelihood that it has already reached the vascular growth phase. Chapter 3 deals with this situation by developing a mathematical model exploring the process that enables the tumor to "soften" the extracellular matrix and invade the neighboring tissue. This invasion may eventually lead to cancer metastasis. Chapter 4 is concerned with the interaction between tumor cells and immune cells. It develops a model of tumor growth which includes tumor-infiltration cytotoxic lymphocytes in a relatively small tumor prior to tumor-induced angiogenesis. The models in Chaps. 2–4 are based on partial differential equations.

Chapter 5 deals with cancer chemotherapy. Two major obstacles to successful chemotherapy of cancer are cell-cycle phase dependence of treatment, and emergence of resistance of cancer cells to cytotoxic agents. The chapter develops optimal control models with the aim of making chemotherapeutic processes more successful.

Finally, Chap. 6 gives an overview of mathematical models of solid tumors and cancer therapy which developed in the last several decades. Here the emphasis is on novel mathematical problems. The models are generally based on partial differential equations in a domain, the tumor region, which is one of the unknowns of the problem. The chapter presents open problems for mathematicians.

I wish to express my appreciation and thanks to Baltazar Aguda, Howard Levine, Marit Nilsen-Hamilton, Mark Chaplain, Anastasios Matzavinos, Marek Kimmel, Georgios Lolas and Andrzej Swierniak for their marvelous contributions. I hope this volume will serve as a useful introduction to those who want to learn about important and exciting problems that arise in modeling of cell division cycle, cell proliferation, and cancer.

Ohio
September 2005   *Avner Friedman*

Contents

Modeling the Cell Division Cycle

Baltazar D. Aguda

Bioinformatics Institute, 30 Biopolis Street, #07-01 Matrix, Singapore 138671
baltazar@bii.a-star.edu.sg

1 Introduction

The ability of an organism to reproduce and perpetuate its species is one of life's defining attributes. As far as we know of all life forms on earth, the replication of a set of genetic information encoded in the DNA is absolutely required. Because of stringent requirements for faithful gene replication and the uncertainty in environmental conditions, the biological cell – with its semi-permeable membrane delineating the DNA from the surroundings – is a structural necessity. In other words, the cell could be viewed as the unit of life. The cell division cycle, or "cell cycle" for short, is the process of replicating the genetic material as well as other cellular components. In this chapter, the reader is introduced to the current consensus view of the molecular machinery of the cell cycle. The emphasis is not on the details of the biochemistry, but rather on the dynamics arising from the topology of the network of molecular interactions. After a summary of the physiology of the cell cycle, the key regulatory molecules are introduced. Using basic chemical reaction kinetics, the rates of the steps in the mechanism can be written and a set of dynamical equations are established. A detailed discussion of the work on fission yeast by Novak et al. [1] is provided to illustrate the intricacies and problems in modeling the cell cycle. The more complex mammalian cell cycle is also discussed briefly to show the evolutionary conservation of molecular pathways essential in cell cycle regulation.

2 Basic Biology of the Cell Division Cycle: Chromosome Cycle and Growth Cycle

Just prior to cell division, in a phase called "mitosis" or M phase, the dramatic events during the segregation of duplicated chromosomes could be visualized under a microscope as shown in Fig. 1. Before mitosis, the DNA content of the

B.D. Aguda: *Modeling the Cell Division Cycle*, Lect. Notes Math. **1872**, 1–22 (2006)
www.springerlink.com   © Springer-Verlag Berlin Heidelberg 2006

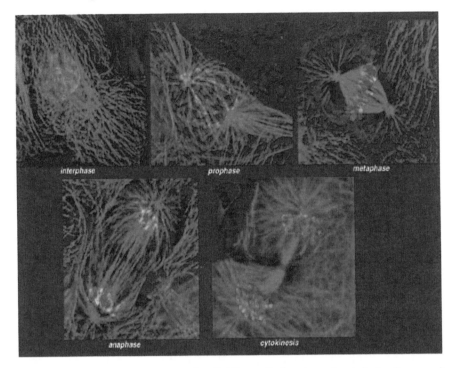

Fig. 1. Fluorescence microscopy of a dividing cell showing microtubules (in green) and chromatin (in red). In "interphase" (composed of the cell cycle phases G1, S, and G2) chromatin is uncondensed and nuclear membrane is intact (not visible here). Chromatin is condensed in "prophase" (when the nuclear membrane also breaks down). Pairs of sister chromatids migrate (with the help of mitotic spindle fibers) to the mid-plane during "metaphase". The sisters are separated during "anaphase" and are pulled towards the spindle poles. In "telophase" (not shown here), nuclear membranes form around the two sets of segregated chromosomes. "Cytokinesis" is the process of division of the two daughter cells. (Photograph courtesy of W. C. Earnshaw, Wellcome Trust Centre for Cell Biology, University of Edinburgh, Scotland, UK)

cell is replicated in S phase (S for $\underline{S}$ynthesis of DNA). A normal "chromosome cycle" involves alternating S and M phases with "$\underline{G}$ap" phases called $G1$ and $G2$. The canonical sequence of eukaryotic cell cycle phases and their typical duration (in mammalian cells) are shown in Fig. 2. In early embryonic cell cycles, S and M phases could alternate without discernible $G1$ and $G2$.

In addition to the chromosome cycle, a cell must also grow and replicate its other components (proteins, membranes, organelles, etc.) and divide these components more or less evenly between the two progeny cells. Doubling of cell mass prior to division is especially important for unicellular organisms such as yeasts in order to maintain a constant average cell size.

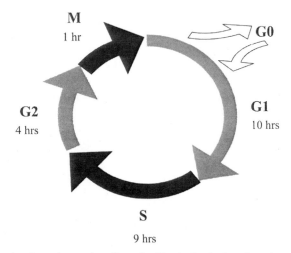

Fig. 2. Phases in the eukaryotic cell cycle. Typical relative durations of the phases are given for a mammalian cell dividing with a period of 24 hours. S = DNA synthesis phase, M = mitosis, G1 and G2 are "gap" phases, G0 = quiescent (nondividing) state

3 The Molecular Regulators of the Cell Cycle: CDKs and APC

In 2001 the Nobel Prize for Physiology or Medicine was awarded to Paul Nurse, Timothy Hunt and Leland Hartwell for their seminal discoveries of key molecular regulators of the cell cycle. Nurse identified and characterized enzymes called "*C*yclin-*D*ependent *K*inases" (CDKs). These enzymes drive the cell cycle by catalyzing phosphorylation of proteins crucial for cell cycle progression. Hunt discovered a group of proteins called "cyclins" that bind CDKs to form cyclin/CDK complexes. The binding of a cyclin is absolutely required for the activation of a CDK. The cyclins are so named because they are degraded periodically during the cell cycle which also explains the observed periodic variation of the kinase activities of CDKs. Hartwell's group discovered many genes involved in cell cycle control, including a set of genes called "start" genes that are crucial for initiating S phase.

Cyclins are synthesized and then abruptly degraded upon exit from mitosis. The enzymatic activities of the CDK partners follow suit. Many of the phosphorylation targets of CDKs have been identified and, although many are still unknown, there is little doubt that the CDKs are the master regulators of various cell cycle-related processes including gene expression, protein degradation (as in nuclear envelop breakdown in mitosis), replication of centrosomes, segregation of duplicated chromosomes, etc. This is why the CDKs are often referred to as components of the "cell cycle engine." In the models presented below, the implicit assumption is that the oscillations in CDK activities drive all other cell cycle processes.

The separation of the duplicate chromosomes (also called sister chromatids) during anaphase requires the activity of an enzyme complex called the "Anaphase-Promoting Complex" (APC) which targets proteins called "cohesins" for degradation (cohesins act as the glue between sister chromatids). It is now known that the APC targets the mitotic cyclins for destruction and, importantly, that this removal of mitotic cyclins is required for exit from mitosis. Indeed, Tyson, Csikasz-Nagy and Novak [2] propose that "to understand the molecular control of cell reproduction is to understand the regulation of CDK and APC activities." Basically, when CDK activity is low (as in G1), APC activity is high; and when CDK activity is high (as in S/G2/M phases), APC activity is low. Thus, one can view the periodic oscillations of CDK activity as resulting from a mutual antagonism between CDK and APC. In the next few paragraphs, a sketch of the essential pathways will be given.

4 The Key Molecular Pathways

Several years of collaboration between the groups of Bela Novak and John Tyson have resulted to a set of models of the eukaryotic cell cycle [1–5]; the essentials of which are shown schematically in Fig. 3 for the particular case of fission yeast (*Schizosaccharomyces pombe*). In this section, an overview of the key regulatory network is given in order to motivate the detailed mathematical modeling that follows.

Two "commitment" points (irreversible transitions) in the cell cycle are shown in Fig. 3. START refers to the point of commitment for another round of DNA replication. FINISH is the point of exit from mitosis where the commitment to divide is made. As suggested in the model discussed below, START and FINISH transitions are associated with the switching on or off of CDK activity, and that these transitions can be viewed as bifurcation points of a dynamical model.

The Novak-Tyson (NT) model depicted in Fig. 3 focuses on the mutual antagonisms between Cdc13/Cdc2 and APC (via the auxiliary factors Ste9 and Slp1), and between Cdc13/Cdc2 and Rum1 (a CDK inhibitor; also referred to as CKI). Cdc13 is a B-type cyclin, and Cdc2 is a cyclin-dependent kinase (now also called CDK1). Binding of Rum1 to Cdc13/Cdc2 inhibits Cdc2 kinase activity (process 1 in Fig. 3). In turn, active Cdc2 can phosphorylate Rum1 which leads to Rum1's degradation (process 2). Ste9/APC modifies Cdc13 to a form that renders this cyclin for degradation (process 3). On the other hand, Cdc13/Cdc2 inhibits Ste9/APC by phosphorylation (process 4). These mutual antagonisms imply that Cdc13/Cdc2 tends not to coexist with either Ste9/APC or Rum1. In this version of the NT model, high CDK and low APC activities are associated with the S/G2/M phases, while low CDK and high APC activities are associated with the G1 phase.

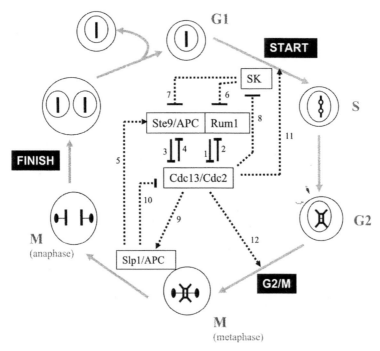

Fig. 3. The fission yeast cell cycle showing the major phase transitions (START, G2/M, and FINISH) along with the network of regulation of the primary cyclin/CDK called Cdc13/Cdc2. See text for discussion of this network. The figure is redrawn and modified from [1]

How are the toggle switches between CDK and APC, and between CDK and CKI, regulated? There are "helper" [1] molecules that assist the CKI and the APC to gain the upper hand over the CDK. As shown in Fig. 3, the helper for Ste9/APC is Slp1/APC (processes 5 and 10). On the other hand, Cdc2 is helped by "starter kinases" (SK) which inactivate Rum1 (process 6) and Ste9/APC (process 7). Cdc13/Cdc2 inhibits SK activity (process 8) by downregulating the latter's transcription (phosphorylation of transcription factors for SK). Process 9 represents experimental observations regarding Cdc2-dependent activation of Slp1/APC. Processes 11 and 12 signify the many phosphorylation processes catalyzed by Cdc2 in order to implement the START and G2/M transitions. To allow the Cdc2 activity to oscillate between low (G1-phase) and high (S/G2/M-phase) values, negative feedback loops exist, namely, the set of processes {9, 5, 3} and the set of processes {8, 6, 1} and {8, 7, 3}. These negative feedback loops allow the possibility of generating oscillatory dynamics.

5 From Qualitative Network and Mechanisms to Dynamical Equations

In this section, mechanistic details of the qualitative network shown in Fig. 3 are discussed. After the rate expressions for the steps in the mechanism are specified, deterministic ordinary differential equations can be written to describe the dynamics of the system. Novak's and Tyson's groups have published various versions of their fission yeast cell cycle model, but for pedagogic reasons, only one of their papers [1] is discussed here in detail to illustrate the formulation of the dynamical equations, computer simulations and bifurcation analysis. For other descriptions and more analysis of the fission yeast cell cycle model, see [2] and [3]. The details of the mechanism that corresponds to Fig. 3 are shown in Fig. 4, and the corresponding dynamical equations are listed in Table 1.

Growth Cycle

Cell mass, M, is modeled by exponential growth ((1), Table 1). M is divided by 2 at the end of mitosis when MPF (mitosis promoting factor) activity decreases

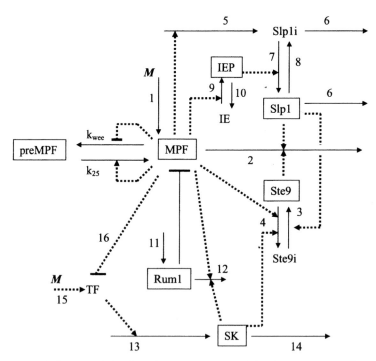

Fig. 4. Detailed steps of the Novak-Tyson model of the fission yeast cell cycle [1]. The numbering of the steps corresponds to the numbering of the rate coefficients used in Table 1 in the text. The figure is redrawn from [1]

Table 1. Equations for the Novak-Tyson (NT) Model for Fission Yeast Cell Cycle (From Table I of [1])

Eqn. No.	Kinetic Equations
1	$\frac{dM}{dt} = \mu M$
2	$\frac{d[Cdc13_T]}{dt} = k_1 M - k_2'[Cdc13_T] - k_2''[Ste9][Cdc13_T] - k_2'''[Slp1][Cdc13_T]$
3	$\frac{d[preMPF]}{dt} = k_{\text{wee}}([Cdc13_T] - [preMPF]) - k_{25}[preMPF]$ $\qquad - k_2'[preMPF] - k_2''[Ste9][preMPF] - k_2'''[Slp1][preMPF]$
4	$\frac{d[Slp1_T]}{dt} = k_5' + \frac{k_5''[MPF]^4}{J_5^4+[MPF]^4} - k_6[Slp1_T]$
5	$\frac{d[Slp1]}{dt} = \frac{k_7[IEP]([Slp1_T]-[Slp1])}{J_7+([Slp1_T]-[Slp1])} - \frac{k_8[Slp1]}{J_8+[Slp1]} - k_6[Slp1]$
6	$\frac{d[IEP]}{dt} = \frac{k_9[MPF](1-[IEP])}{J_9+(1-[IEP])} - \frac{k_{10}[IEP]}{J_{10}+[IEP]}$
7	$\frac{d[Ste9]}{dt} = \frac{(k_3'+k_3''[Slp1])(1-[Ste9])}{J_3+(1-[Ste9])} - \frac{(k_4'[SK]+k_4[MPF])[Ste9]}{J_4+[Ste9]}$
8	$\frac{d[Rum1_T]}{dt} = k_{11} - (k_{12} + k_{12}'[SK] + k_{12}''[MPF])[Rum1_T]$
9	$\frac{d[SK]}{dt} = k_{13}[TF] - k_{14}[SK]$

Auxiliary Equations

i	$k_{\text{wee}} = k_{\text{wee}}' + (k_{\text{wee}}'' - k_{\text{wee}}')G(V_{\text{awee}}, V_{\text{iwee}}[MPF], J_{\text{awee}}, J_{\text{iwee}})$
ii	$G(a,b,c,d) = \frac{2ad}{b-a+bc+ad+\sqrt{(b-a+bc+ad)^2-4ad(b-a)}}$
iii	$k_{25} = k_{\text{wee}}' + (k_{25}'' - k_{25}')G(V_{a25}[MPF], V_{i25}, J_{a25}, J_{i25})$
iv	$[Trimer] = \frac{2[Cdc13_T][Rum1_T]}{\Sigma+\sqrt{\Sigma^2-4[Cdc13_T][Rum1_T]}}$
v	$\Sigma = [Cdc13_T] + [Rum1_T] + K_{\text{diss}}$
vi	$[MPF] = \frac{([Cdc13_T]-[preMPF])([Cdc13_T]-[Trimer])}{[Cdc13_T]}$
vii	$[TF] = G(k_{15}M, k_{16}' + k_{16}''[MPF], J_{15}, J_{16})$

below 0.1, a threshold value chosen by Novak et al. [1] to simulate experiments as closely as possible. The lack of explicit mechanistic coupling between mass and the activation of CDK activity is a major shortcoming of models of the cell cycle at this time. In the NT model, mass is coupled with $[Cdc13_T]$ ((2), Table 1) and with [TF] (equation vii, Table 1; TF = transcription factor). Note that the use of the Goldbeter-Koshland [6] function G for [TF] allows for an ultrasensitive response of [TF] to increase in mass. This function is defined in equation (ii) of Table 1. More detailed explanation about the G function is given in the Appendix.

MPF Dynamics

In Table 1, (2), Cdc13$_T$ refers to the total of all cdc13/cdc2 complexes, including Rum1/cdc13/cdc2 trimers. The first term on the right-hand side (r.h.s.) of (2) assumes that the rate of synthesis of Cdc13 is proportional to the cell's mass. The succeeding terms are degradation rates, the first being due to degradation pathways that are independent of Ste9 and Slp1, while the last two terms are due to Ste9- and Slp1-induced degradation, respectively.

Figure 5 shows the interconversion of the various species containing Cdc13/Cdc2. The species called preMPF refers to the dimeric and trimeric tyrosine-phosphorylated forms of Cdc13/Cdc2. From this figure, one can understand the equation for preMPF ((3), Table 1). The first term on the r.h.s. gives the rate of formation of preMPF due to steps catalyzed by Wee1. The second term gives the rate of dephosphorylation of preMPF catalyzed by Cdc25. The last 3 terms are degradation terms similar to those of Cdc13$_T$. Note that

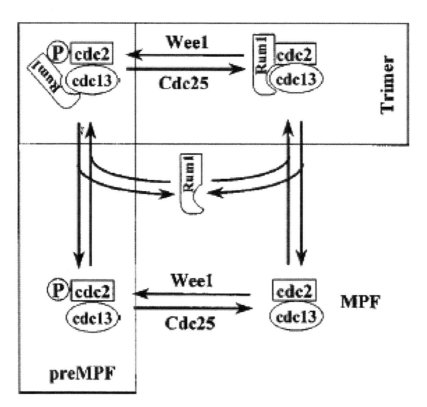

Fig. 5. Various protein complexes involving Cdc13/Cdc2, including the trimers with Rum1 (tyrosine-phosphorylated and unphosphorylated). preMPF is defined as the total of the tyrosine-phosphorylated Cdc13/Cdc2, with and without Rum1. The figure is from [1] and used here with permission

k_{wee} is assumed to have the form given in equation (i) in Table 1. The second term is proportional to a G function whose second argument is a function of [MPF]. Increasing [MPF] therefore decreases k_{wee} (because G increases with the ratio $V_{\text{awee}}/(V_{\text{iwee}}[\text{MPF}]))$; thus, the form of k_{wee} encapsulates the mutual antagonism between Wee1 and MPF. The first term on the r.h.s. of the k_{wee} equation represents MPF-independent phosphorylation of Wee1.

An explanation similar to that of k_{wee} can be given for the equation for k_{25} (equation iii, Table 1). Note the dependence of k_{25} on [MPF]. The rate parameter k_{25} increases with [MPF] since G increases with the ratio $V_{\text{a25}}[\text{MPF}]/V_{\text{i25}}$). This dependence represents an important positive feedback loop between Cdc25 and MPF [7].

The equation for [Trimer], (iv) and (v) in Table 1, assumes rapid equilibrium between trimers and the components cdc13/cdc2 and Rum1. The equation for active MPF is given in (vi) in Table 1.

Dynamics of the FINISH Module (Slp1/APC and IEP)

The total Slp1 concentration, $[\text{Slp1}_T]$, varies according to (4) of Table 1. The first term is due to MPF-independent synthesis of Slp1. The second term is MPF-dependent synthesis of Slp1 which is a Hill-type equation with high cooperativity as represented by the exponent 4 of [MPF]. This provides a switch-like behavior for Slp1 that is controlled by MPF. The last term on the r.h.s. represents degradation.

Active Slp1 varies according to (5) in Table 1. The interpretation of this equation could be readily understood from Fig. 4. The two enzymatic steps (7 and 8 of Fig. 4) have Michaelis-Menten kinetics. The last term represents degradation.

The hypothetical enzyme in its phosphorylated form, namely IEP, varies according to (6) in Table 1. Novak and Tyson predict the existence of IEP which provides the delay necessary to give sufficient time for the chromosomes to align with the metaphase plane before they are separated at anaphase (Slp1/APC catalyzing the degradation of cohesins). Figure 4 gives a straightforward interpretation of the terms on the r.h.s. of (6) in Table 1. Note that [IE] + [IEP] = 1.

G1/S Module

Ste9, the other APC auxiliary factor involved mainly in the G1/S module, varies according to (7) in Table 1. The first term on the r.h.s. represents the activation of the inactive form of Ste9 (for both Slp1-independent and Slp1-dependent activation). The second term is for the deactivation of Ste9 through SK-catalyzed and MPF-catalyzed steps.

The total concentration $[\text{Rum1}_T]$ varies according to (8) in Table 1. The first term is for the assumed constant synthesis rate of the Rum1 protein. The second term represents the three pathways of Rum1 degradation, including SK-dependent and MPF-dependent routes.

The starter kinase SK concentration varies according to (9) in Table 1. Note that mass M drives the increase in TF activity according to the G function discussed previously.

Computer Simulations

Solving the differential equations given in Table 1 gives the periodic oscillations shown in Fig. 6A. The parameters used in this figure are assumed by Novak et al. [1] to correspond to "wild-type" fission yeast. The dimensionless cell mass M increases from 1 to 2, and it should be noted that the "periodicity" in the growth cycle is actually imposed by dividing M by 2 at the end of mitosis when MPF decreases through the value of 0.1.

Note that cells enter mitosis when MPF activity shoots up and, after a time delay provided by IEP, Slp1/APC gets activated and initiates the degradation of Cdc13 (steps 9, 7, 2 in Fig. 4) with a corresponding drop in MPF activity. Ste9/APC then gets activated (step 3 in Fig. 4) and helps in the accelerated degradation of Cdc13 causing the FINISH transition. Thus, a new cycle begins with very low MPF activity and active TF which induces expression of SK (step 13 in Fig. 4). The rapid increase in SK activity will inhibit Ste9 and Rum1, thus allowing the moderate increase in MPF and the associated G1/S transition. The G2/M transition characterized by the large and abrupt increase in MPF activity is basically due to the inactivation of Wee1 – carried out by the positive feedback loop between Cdc25 and MPF.

Novak et al. [1] used the model summarized in Table 1 to simulate various mutants, including those missing in wee1, rum1, ste9 and cdc25. The reader is referred to [1] for more details.

Phase-Plane and Bifurcation Analysis

Although the NT model has 9 independent dynamical variables, it is interesting to see how Novak et al. [1] carried out a phase-plane analysis in order to develop an intuitive understanding of the cell cycle transitions. The authors determined the nullclines on the $[Cdc13_T]$-$[MPF]$ plane shown in Figs. 6B and 6C. The $[Cdc13_T]$-nullcline is determined by setting $d[Cdc13_T]/dt$ to zero ((2) of Table 1) and obtaining an expression containing $[Cdc13_T]$ and $[MPF]$ only. The mass M in this nullcline function is considered a parameter. The other variables, $[Ste9]$ and $[Slp1]$, are ultimately expressed in terms of mass M and $[MPF]$ after setting all the other variables at steady state (see [1] for details). The $[MPF]$-nullcline function is derived from equation vi of Table 1 after solving for the steady state values of $[preMPF]$ and $[Trimer]$.

Figures 6B and 6C show two phase planes, one for mass $M = 1$ (newborn cells) and for $M = 1.6$ (cells that just passed the G2/M transition). M is treated as a control parameter. For newborn cells, the two nullclines intersect

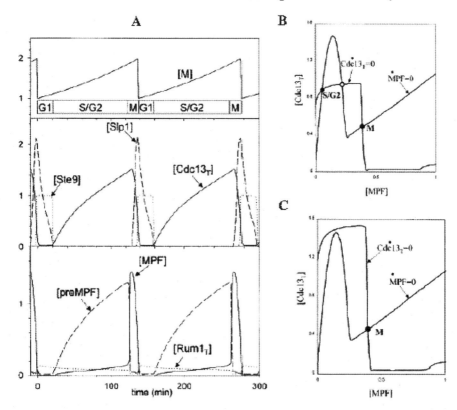

Fig. 6. (**A**) Numerical simulations of the Novak-Tyson model defined in Table 1. Figure is copied from [1], and used here with permission. The parameter values are those from Table II of the same reference and listed here for convenience (all constants have units of min^{-1} except the J_i's and k_{diss} which are dimensionless): $k_1 = 0.03$, $k_2' = 0.03$, $k_2'' = 1$, $k_2''' = 0.1$, $k_3' = 1$, $k_3'' = 10$, $J_3 = 0.01$, $k_4' = 2$, $k_4 = 35$, $J_4 = 0.01$, $k_5' = 0.005$, $k_5'' = 0.3$, $k_6 = 0.1$, $J_5 = 0.3$, $k_7 = 1$, $k_8 = 0.25$, $J_7 = J_8 = 0.001$, $k_9 = 0.1$, $k_{10} = 0.04$, $J_9 = J_{10} = 0.01$, $k_{11} = 0.1$, $k_{12} = 0.01$, $k_{12}' = 1$, $k12'' = 3$, $k_{diss} = 0.001$, $k_{13} = 0.1$, $k_{14} = 0.1$, $k_{15} = 1.5$, $k_{16}' = 1$, $k_{16}'' = 2$, $J_{15} = J_{16} = 0.01$, $V_{awee} = 0.25$, $V_{iwee} = 1$, $J_{awee} = J_{iwee} = 0.01$, $V_{a25} = 1$, $V_{i25} = 0.25$, $J_{a25} = J_{i25} = 0.01$, $k_{wee}' = 0.15$, $k_{wee}'' = 1.3$, $k_{25}' = 0.05$, $k_{25}'' = 5$, $\mu = 0.005$. (**B**) Cdc13$_T$ and MPF nullclines for M = 1 (newborn cells), and (**C**) for M = 1.6 (cells that just passed the G2/M transition). • = stable steady state; o = unstable steady state. Figure is from [1] with permission

3 times, giving rise to three steady states, two stable (solid circles) and one unstable (open circle). The S/G2 cell cycle phase is associated with high [$Cdc13_T$] level but low [MPF] activity. Upon reaching a mass threshold, the S/G2 state and the unstable steady state are lost, and the system switches to a steady state characterized by lower [$Cdc13_T$] but higher [MPF] activity – this state is associated with M-phase. Note that the steady-state associated with M-phase is stable in the phase plane of Fig. 6C; when the full set of

dynamical equations is considered (e.g. including M as a dynamical variable), the M-phase steady state becomes unstable allowing a FINISH transition into G1.

6 Mammalian Cell Cycle and Checkpoints

The knowledge learned from lower eukaryotes such as yeast is quite useful in unraveling the more complex regulatory network of the mammalian cell cycle. There are at least a dozen more types of CDKs and cyclins in mammals, [8–9] and particular cyclin/CDK complexes seem to exert their functions at specific phases of the cell cycle. The current consensus picture is as follows: D-type cyclins are expressed early in G1; these cyclins bind and activate CDK4 and/or CDK6. The D-type cyclins are sometimes referred to as "growth-factor sensors". Prior to S-phase, a G1 checkpoint reminiscent of START in yeast, but called "Restriction Point" (or R-point) in mammals, is crossed via the activation of cyclin E/CDK2 [8, 10–11]. The maintenance of S-phase requires the activities of cyclin A/CDK2 and cyclin A/CDK1 [9]. The G2/M transition is triggered by an exponential increase in the activity of cyclin B/CDK1 (also called MPF or "mitosis promoting factor" in the older literature). As in yeast, the degradation of cyclin B is required for exit from mitosis. There had been attempts at modeling the mammalian cell cycle [8, 12–14]. The discussion here will focus on the control mechanism of the G1/S transition guarded by the R-point. The suggestion will also be made that the G2/M transition is basically controlled in the same way, i.e. that there is an intrinsic instability in the network which is targeted by signaling pathways to arrest or slow down cell cycle progression. These ideas are based on the results of Aguda and co-workers [7–8, 15–17].

A qualitative network of the G1/S transition is shown in Fig. 7. The pre-replication complex (pre-RC) is composed of several protein complexes (ORC, MCMs, Cdc6, etc.) involved in recognition of origins and initiation of DNA replication (e.g. opening the DNA duplex). Transcription factors of the E2F family have been shown to induce expression of various S-phase genes, including many members of the pre-RC, cyclin E, cyclin A, Cdc25A, etc. [18]. In quiescent cells, the E2Fs are inhibited by members of the retinoblastoma protein (RB) family. This inhibition is due to binding of RB to E2F, as well as recruitment by RB of HDAC proteins that transform chromatin structure into one that is repressesive of transcription [19]. Upon growth factor (GF) stimulation, D-type cyclins start the cascade of phosphorylation steps leading to inactivation of RB. Active CDK4 or CDK6 phosphorylates RB causing the latter's dissociation from E2F; cyclin E and cyclin A can then be expressed and subsequently activate CDK2 which further phosphorylates RB. The positive feedback loops in RB phosphorylation (see Fig. 7) are obvious candidates for the explanation of the switch-like behavior of the R-point.

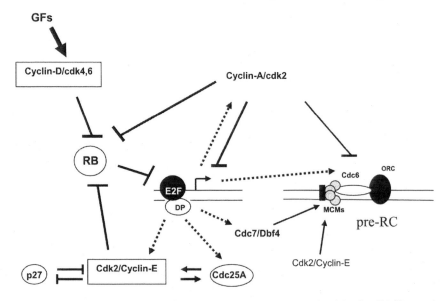

Fig. 7. A qualitative network of the major interactions involved in the G1/S transition in the mammalian cell cycle. Arrows mean "activate" and hammerheads mean "inhibit". *Dashed* lines are transcriptional and translational processes. *Solid* lines are protein-level interactions. See text for details

In addition to RB, there are CDK Inhibitor proteins (collectively called "CKIs") that bind and inhibit the CDKs. To simplify the discussion, only one of them is shown in Fig. 7, namely, the CKI called p27 (also named Kip1). This protein has been shown to play a significant role in regulating the G1/S transition via a mutual antagonistic relationship with CDK2. Also, as shown in Fig. 7, a positive feedback loop between the phosphatase Cdc25A and CDK2 exists. Note that this Cdc25A-CDK2 loop is similar to the positive feedback represented in the parameter k_{25} for the G2/M transition in the Novak-Tyson model (in mammals, the isoform Cdc25C is involved in a positive feedback loop with CDK1 during the G2/M transition). The hypothesis that have been advanced previously [8] regarding the "kinetic origin" of the switching behavior of the R-point is that the subnetwork involving CDK2, Cdc25A, and p27Kip1 guarantees the existence and sharpness of the activation switch of CDK2. The CDK2-Cdc25A-p27 subnetwork is characterized by two coupled positive loops that generate the sharpness of the CDK2 switch.

The details of the positive feedback loop between CDK2 and Cdc25A is interesting. As shown in Fig. 8, both proteins undergo phosphorylation-dephosphorylation (PD) cycles. Active Cdc25A removes an inhibitory phosphate group from CDK2. In return, active CDK2 phosphorylates and thereby activates Cdc25A. It was shown previously [7] that several interactions

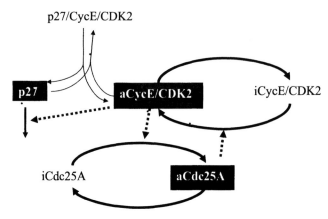

Fig. 8. Mechanistic details of the CDK2-Cdc25A-p27Kip1 subnetwork. Inactive Cdc25A (iCdc25A) is transformed to the active state (aCdc25A) by its phosphorylation by active Cyclin E/CDK2 (aCycE/CDK2). Inactive Cyclin E/CDK2 (iCycE/CDK2) is dephosphorylated by aCdc25A and transformed to active Cyclin E/CDK2 (aCycE/CDK2). The inhibitor p27Kip1 forms a trimer with CycE/CDK2 thereby repressing the kinase activity of CDK2; aCycE/CDK2 phosphorylates p27Kip1 rendering the latter a target of the protein degradation machinery

between cyclic enzyme reactions (PD cycles being specific examples) are unstable. For example, positively coupled PD cycles exhibit transcritical bifurcation as shown in Fig. 9. For the specific example of mass-action kinetics assumed in this figure, a set of independent dynamical equations is the following:

$$\frac{dy_1}{dt} = v_{1f} - v_{1r}$$

$$\frac{dy_2}{dt} = v_{2f} - v_{2r}$$

(1)

where $v_{1f} = k_{1f}x_1y_2$, $v_{1r} = k_{1r}y_1$, $v_{2f} = k_{2f}x_2y_1$, and $v_{2r} = k_{2r}y_2$. Note that there are two conservation conditions, $E_1 = x_1 + y_1$ and $E_2 = x_2 + y_2$.

There are two branches of steady states, one is a zero steady state for all total enzyme concentrations while the other is a branch with a positive slope. The transcritical bifurcation point, where these two branches intersect, is determined from the following equation (for mass-action kinetics):

$$E_1 E_2 = \frac{k_{1r}}{k_{1f}} \frac{k_{2r}}{k_{2f}}$$

(2)

It is interesting to notice from the above equation and from Fig. 9 that in order for Y_1 and Y_2 to have positive steady states, the product $E_1 E_2$ must be greater than the value of the r.h.s. of (2). If Y_1 and Y_2 are interpreted as the active states of proteins 1 and 2, one can then claim that these proteins must increase beyond threshold levels defined by (2). This is one possible mechanism

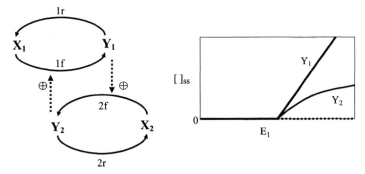

Fig. 9. Transcritical bifurcation diagram generated by positively coupled cycles. E_1 is the sum of $[X_1]$ and $[Y_1]$. The steady state concentrations ($[\ ]_{ss}$) of Y_1 and Y_2 are plotted versus E_1. *Solid* curves are stable steady states; *dashed* line is a branch of unstable steady states

that may be relevant to the important issue of coupling cell growth with the cell cycle.

For the G2/M transition, a subnetwork topology similar to the CDK2-Cdc25A-p27 subnetwork in G1/S can be identified. This is the CDK1-Cdc25C-Wee1 that is also a component of the Novak-Tyson model. Indeed, it has been suggested that the CDK1-Cdc25C-Wee1 subnetwork is a target of DNA damage signal transduction pathways to arrest the cell cycle [15].

7 Coupling Between the Cell Cycle and Apoptosis

When the cell cycle machinery is in overdrive (e.g. due to overexpression of cancer-causing genes called oncogenes or, alternatively, due to inhibition of tumor suppressor genes), cells may "commit suicide" or apoptosis for the well being of the multicellular organism. Aguda & Algar [17] recently reviewed and analyzed the complex molecular networks linking the initiation of S phase and the activation of enzymes called caspases that execute the breakdown of cellular proteins during apoptosis. These authors also proposed a modularization scheme (Fig. 10A) that subdivides the network into modules corresponding to signal transduction pathways, S-phase initiation, apoptosis, and a control node that coordinates the cell cycle and apoptosis. An example of a kinetic model for this modular network is shown in Fig. 10B. Craciun, Aguda & Friedman [20] carried out a detailed mathematical analysis of the Aguda-Algar model using the following set of equations:

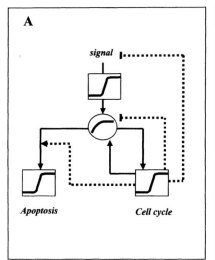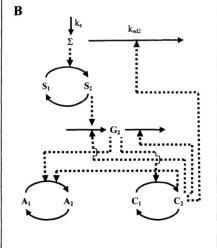

Fig. 10. (A) The modular structure of the Aguda-Algar model that links the initiation of the cell cycle and apoptosis with growth-factor signaling [17]. **(B)** A kinetic model corresponding to (A) that implements the ultrasensitive responses of the signaling, cell cycle, and apoptosis modules. The cyclic enzyme reactions involving the couples (S_1, S_2), (C_1, C_2), and (A_1, A_2) possess the zeroth-order ultrasensitivity discussed in the Appendix

$$\frac{d\Sigma}{dt} = \varepsilon^0(k_s - k_{sd1}\Sigma - k_{sd2}C_2\Sigma) \tag{3}$$

$$\frac{dS_2}{dt} = \varepsilon^{-2}\left(\frac{k_1\Sigma S_1}{K_{M1} + S_1} - \frac{v_{m1}S_2}{K_{Mr1} + S_2}\right) \tag{4}$$

$$\frac{dG_2}{dt} = \varepsilon^{-3}(k_2S_2 + k_{2a}C_2 - k_{m2}G_2 - k_{m2a}C_2G_2) \tag{5}$$

$$\frac{dC_2}{dt} = \varepsilon^{-1}\left(\frac{k_3G_2C_1}{K_{M3} + C_1} - \frac{v_{m3}C_2}{K_{Mr3} + C_2}\right) \tag{6}$$

$$\frac{dA_2}{dt} = \varepsilon^2\left(\frac{k_4G_2A_1}{K_{M4} + A_1} + \frac{k_{4a}C_2A_1}{K_{M4} + A_1} - \frac{v_{m4}A_2}{K_{Mr4} + A_2}\right) \tag{7}$$

where Σ = signaling molecule, S_2 = active signaling protein, G_2 = control node of transcription factors, C_2 = active cell cycle marker (for initiation of S phase), and A_2 = active apoptosis marker. S_1, C_1, and A_1 are the corresponding inactive forms of the molecules.

Using (3)–(7), a phase diagram can be constructed that demonstrates the capability of the modular network (Fig. 10) to simulate a common experimental observation, namely, that increasing intensity of growth factor stimulation (represented by the parameter k_s) can drive the cell from quiescence (nondividing) to cell cycling and ultimately to apoptosis (see [17] for review). A phase diagram is shown in Fig. 11 using the parameters k_s and k_{sd2}; the latter

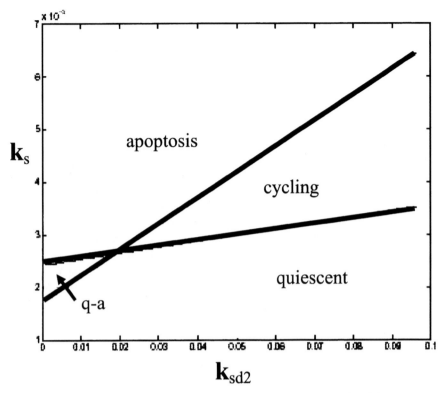

Fig. 11. A phase diagram constructed from the model equations (3)–(7) in the text (from Craciun, Aguda & Friedman [20]). The parameter k_s is associated with the intensity of growth-factor signaling, and the parameter k_{sd2} is associated with the strength of the negative feedback loop from the cell cycle to the signal molecule Σ (see Fig. 10). Fixed parameter values are: $k_{sd1} = 0.03$, $k_1 = k_3 = 10$, $V_{m1} = 1$, $k_{M1} = k_{Mr1} = 0.02$, $k_2 = 0.1$, $k_{2a} = 0.01$, $k_{m2} = 1$, $k_{m2a} = 0.001$, $V_{m3} = 0.1$, $k_{M3} = 0.02$, $k_{Mr3} = 0.02$, $k_4 = 1$, $k_{4a} = 1$, $V_{m4} = 1$, $k_{M4} = 0.02$, $k_{Mr4} = 0.1$, $\varepsilon = 0.1$, $S_1 + S_2 = 1$, $C_1 + C_2 = 1$, $A_1 + A_2 = 1$

parameter is a rate constant associated with the C_2-dependent decay of the signal Σ (i.e. a negative feedback loop from the cell cycle to growth-factor signaling). An interesting prediction from the phase diagram is the region labeled "q-a" where the system is bistable, i.e. the system can either be in a quiescent (q) state or in the apoptotic (a) state depending on its history (or initial conditions).

8 Conclusions and Future Directions

Novak and Tyson have integrated the details of the eukaryotic cell cycle into a coherent picture of a network of molecular interactions whose dynamics could be modeled by ODEs and analyzed using the tools of dynamical systems and bifurcation theories. An implicit assumption in the NT model is that there is an autonomous oscillator, namely the CDK oscillator, that drives the cell cycle processes (directly or indirectly) of DNA synthesis, centrosome segregation, chromosome segregation, nuclear envelop breakdown, cytokinesis, etc. The NT model further hypothesizes that transitions between cell cycle phases are represented by bifurcations of the model dynamical system, and that cell cycle checkpoints are controlled by the parameters affecting these bifurcations.

Topological analysis of the cell cycle network identified essential positive and negative feedback loops. The switch from G1 to S/G2/M in the NT model is associated with bistability caused by positive loops in the model. The CDK oscillations are possible due to the presence of negative loops. The coupling between these positive and negative loops gives rise to hysteretic oscillations (periodic switching between the lower and upper branches of stable steady states). Breaking down the complex network into these essential loops provides a good way of reducing the model without removing required qualitative features [17]; such a model reduction also leads to the interesting hypothesis that there exists a minimal cell cycle network structure that is conserved from lower to higher eukaryotes, but modified only according to new demands of multicellularity [21].

Future mathematical models of the cell cycle are expected to utilize genomic, transcriptomic, and proteomic data that are rapidly accumulating as a result of recent advances in high-throughput biotechnologies [22–26]. Bioinformaticians are currently creating pathways databases or knowledgebases that integrate heterogeneous sources of information (a good example of a knowledgebase can be found at http://reactome.org). A challenge for mathematicians and modelers will be to devise methods of extracting kinetic models from these pathways databases [27].

References

1. B. Novak, Z. Pataki, A. Ciliberto, and J. J. Tyson (2001) "Mathematical model of the cell division cycle of fission yeast," *Chaos* 11: 277–286.
2. J. J. Tyson, A. Csikasz-Nagy, and B. Novak, "The dynamics of cell cycle regulation," *BioEssays* 24: 1095–1109 (2002).
3. A. Sveiczer, A. Csikasz-Nagy, B. Gyorffy, J. J. Tyson, and B. Novak, "Modeling the fission yeast cell cycle: Quantized cycle times in wee1 *cdcd25Δ* mutant cells," *Proc. Natl. Acad. Sci. USA* 97: 7865–7870 (2000).
4. K. C. Chen, A. Csikasz-Nagy, B. Gyorffy, J. Val, B. Novak, and J. J. Tyson, "Kinetic analysis of a molecular model of the budding yeast cell cycle," *Molecular Biology of the Cell* 11: 369–391 (2000).
5. K. C. Chen, L. Calzone, A. Csikasz-Nagy, F. R. Cross, B. Novak, and J. J. Tyson, "Integrative analysis of cell cycle control in budding yeast," *Molecular Biology of the Cell* 15: 3841–3862 (2004).

6. A. Goldbeter and D. E. Koshland, Jr., "An amplified sensitivity arising from covalent modification in biological systems," *Proc. Natl. Acad. Sci. USA* 78: 6840–6844 (1981).

7. B.D. Aguda, "Instabilities in Phosphorylation-Dephosphorylation Cascades and Cell cycle Checkpoints", *Oncogene* 18: 2846–2851 (1999).

8. B.D. Aguda & Y. Tang, "The kinetic origins of the restriction point in the mammalian cell cycle", *Cell Proliferation* 32: 321–335 (1999).

9. J. Pines, "Four-dimensional control of the cell cycle," *Nat Cell Biol.* 1: E73–E79 (1999).

10. A. B. Pardee, "A restriction point for control of normal animal cell proliferation," *Proc Natl Acad Sci USA.* 71:1286–1290 (1974).

11. S. V. Ekholm, P. Zickert, S. I. Reed, and A. Zetterberg, "Accumulation of cyclin E is not a prerequisite for passage through the restriction point," *Mol Cell Biol.* 21:3256–3265 (2001).

12. V. Hatzimanikatis, K. H. Lee, and J. E. Bailey, "A mathematical description of regulation of the G1-S transition of the mammalian cell cycle," *Biotechnol Bioeng.* 65:631–637 (1999).

13. M. N. Obeyesekere, E. S. Knudsen, J. Y. Wang, and S. O. Zimmerman, "A mathematical model of the regulation of the G1 phase of Rb+/+ and Rb-/- mouse embryonic fibroblasts and an osteosarcoma cell line," *Cell Prolif.* 30:171–194 (1997).

14. Z. Qu, J. N. Weiss, and W. R. MacLellan, "Regulation of the mammalian cell cycle: a model of the G1-to-S transition," *Am J Physiol Cell Physiol* 284: C349–C364 (2003).

15. B.D. Aguda, "A quantitative analysis of the kinetics of the G2 DNA damage checkpoint system", *Proc. Natl. Acad. Sci. USA* 96: 11352–11357 (1999).

16. B.D. Aguda, "Kick-starting the cell cycle: From growth-factor stimulation to initiation of DNA replication", *Chaos* 11: 269–276 (2001).

17. B.D. Aguda & C.K. Algar, "Structural analysis of the qualitative networks regulating the cell cycle and apoptosis", *Cell Cycle* 2: 538–544 (2003).

18. H. Muller and K. Helin, "The E2F transcription factors: key regulators of cell proliferation," *Biochim Biophys Acta* 1470:M1–12 (2000).

19. H. S. Zhang, M. Gavin, A. Dahiya, A. A. Postigo, D. Ma, R. X. Luo, J. W. Harbour, and D. C. Dean, "Exit from G1 and S phase of the cell cycle is regulated by repressor complexes containing HDAC-Rb-hSWI/SNF and Rb-hSWI/SNF," *Cell* 101:79–89 (2000).

20. G. Craciun, B.D. Aguda, and A. Friedman, "A detailed mathematical analysis of a modular network coordinating the cell cycle and apoptosis," Mathematical Biosciences & Eng 2: 473–485 (2005).

21. B. Novak, A. Csikasz-Nagy, B. Gyorffy, K. Nasmyth, J. J. Tyson, "Model scenarios for evolution of the eukaryotic cell cycle," *Philos Trans R Soc Lond B Biol Sci.* 353:2063–2076 (1998).

22. R. J. Cho et al., "A genome-wide transcriptional analysis of the mitotic cell cycle," *Mol Cell* 2: 65–73 (1998).

23. R. J. Cho et al., "Transcriptional regulation and function during the human cell cycle," *Nat Genet* 27: 48–54 (2001).

24. G. Rustici et al., "Periodic gene expression program of the fission yeast cell cycle," *Nat Genet* 36: 809–817 (2004).

25. P. T. Spellman et al., "Comprehensive identification of cell cycle-regulated genes of the yeast Saccharomyces cerevisiae by microarray hybridization," Mol Biol Cell 9: 3273–3297 (1998).
26. M. Whitfield et al., "Identification of genes periodically expressed in the human cell cycle and their expression in tumors," *Molecular Biology of the Cell* 13: 1977–2000 (2002).
27. B.D. Aguda, G. Craciun, R. Cetin-Atalay, "Data sources and computational approached for modeling gene networks," *Reviews in Computational Chemistry,* 21: 381–411 (2005).

Appendix

Zeroth-Order Ultrasensitivity and the Goldbeter-Koshland Function

The Goldbeter-Koshland function G is used in the Novak-Tyson (NT) model to generate a switching behavior when certain parameters are varied. Here, it is shown how this switching behavior called zeroth-order ultrasensitivity [6] arises when the Michaelis constants (J_m) are small, and what parameters can be varied to see this ultrasensitivity. A very small J_m means that the enzyme-substrate complex is very tight and hardly dissociates. Thus, the NT model that uses the function G makes the assumption that the enzyme-substrate complexes involved in the steps are very stable.

For a reversible reaction such as that shown in Fig. 12, if both steps a and i are enzyme-catalyzed steps with Michaelis-Menten kinetics, the rate of change of species A concentration is:

$$\frac{d[A]}{dt} = \frac{V_a[I]}{J_a + [I]} - \frac{V_i[A]}{J_i + [A]} \tag{A.1}$$

Since the mechanism is cyclic, and no mass is lost, the following conservation condition holds

$$[I] + [A] = C, \quad \text{constant} . \tag{A.2}$$

If $[A]$ is at steady state, i.e. $d[A]/dt = 0$, and using the conservation condition to express $[I]$ in terms of $[A]$, one can derive the following equation for the

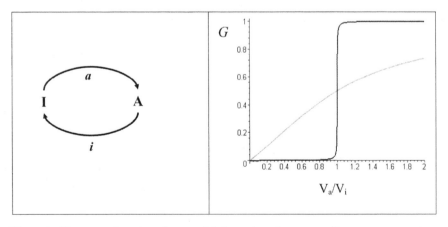

Fig. 12. How zeroth-order ultrasensitivity arises from a cyclic enzyme reaction. Both reaction steps a and i are catalyzed with Michaelis-Menten kinetics, and with maximum steady state rates V_a and V_i, respectively. The Goldbeter-Koshland function, G, is the fraction of species A at steady state. Black curve corresponds to Michaelis constants $J_a = J_i = 0.001$; gray curve is for $J_a = J_i = 1$

fraction $a_s = [A]_s/C$ at steady state (this fraction is none other than the function G). Note that the steady state equation is a quadratic polynomial in a_s and the expression below considers one of the roots (see [4]).

$$a_s = \frac{[A]}{C} = G(V_a, V_i, J_a, J_i) = \frac{2V_aJ_i}{B + \sqrt{B^2 - 4V_aJ_i(V_i - V_a)}} \qquad \text{(A.3)}$$

where

$$B = V_i - V_a + V_i\frac{J_a}{C} + V_a\frac{J_i}{C}$$

A plot of G versus (V_a/V_i) for various values of Michaelis constants, J_a and J_i, is shown in Fig. 12. Note that as the J's go to zero, G becomes a step function.

Angiogenesis-A Biochemical/Mathematical Perspective

Howard A. Levine[1] and Marit Nilsen-Hamilton[2]

[1] Department of Mathematics, Iowa State University, Ames, IA 5011
[2] Department of Biochemistry, Biophysics & Molecular Biology, Ames, IA 5011

Abstract. Angiogenesis, the formation of new blood vessels from an existing vasculature, is a complex biochemical process involving many different biomolecular and cellular players. We propose a new model for this process based upon bio-molecular considerations and the cell cycle.

Key words: angiogenesis, chemotaxis

1 Introduction

Angiogenesis is one of the many important processes that occurs in both normal development such as placental growth and embryonic development as well as in abnormal growth such as in the rapid growth of malignant tumors. The purpose of this paper is to give the mathematically inclined reader a sense of the underlying biochemical and mathematical ideas that have been recently coupled together to model this process. Most of what we have to say here has already appeared in the mathematical biology literature. However, we have added included some new modifications that have yet to appear but that may offer us new insight into modeling this complex phenomenon.

Outline

H.A. Levine and M. Nilsen-Hamilton: *Angiogenesis-A Biochemical/Mathematical Prospective*,
Lect. Notes Math. **1872**, 23–76 (2006)
www.springerlink.com

2 What is Angiogenesis?

The vasculature is the system of channels through the bodies of plants and animals by which proteins and nutrients are distributed to all component cells. In mammals and other vertebrates these component parts of the vasculature take the form of the cardiovascular and lymphatic vascular systems. The cardiovascular system, with its faster circulation rates, is the primary means of delivering oxygen and nutrients to tissues in the mammalian body (Fig. 1). The blood vessels transport a variety of cells such as erythrocytes (red blood cells) and many different cell types of the immune system.

The lymphatic system also provides a means for cells of the immune system to move around the body that is exploited by cancer cells that have become mobile (metastatic). There are two stages of formation of the vasculature. The first occurs *de novo* during embryogenesis and involves a process called vasculogenesis. The lymphatic system is formed later during embryogenesis than the cardiovasculature by a similar process called lymphangiogenesis. The second stage of formation of the vasculature occurs in the adult as new blood or lymphatic vessels sprout from the old. In this way, new capillaries are formed

ANGIOGENESIS

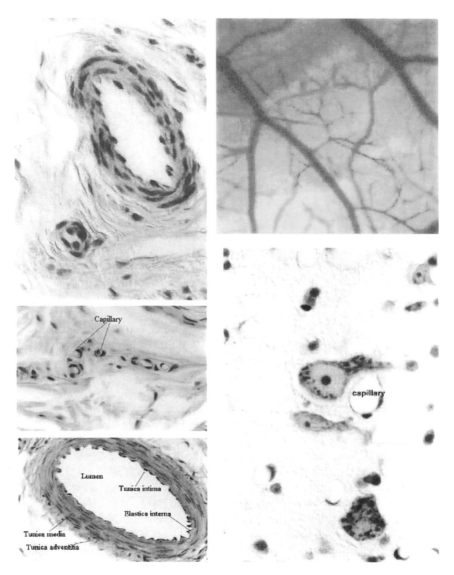

Fig. 1. Aspects of Angiogenesis: Cross sections of capillaries, veins and arteries at various scales as well as the branched network of the vasculature (subfigure at the upper right hand corner). http://cellbio.utmb.edu/microanatomy/ cardiovascular/cardiovascular_system.htm

that penetrate new tissues that form during processes such as in wound heal-ing and mammary, uterine and placental growth during reproduction. The body's ability to initiate angiogenesis and send capillaries into new tissues is also used to advantage by tumors to supply their nutritional needs. Endothe-lial cells (EC) form the linings of the vertebrate vasculatures. The channel (lumen) of the blood or lymphatic vessel is enclosed by ECs that are sealed, one to the other at their periphery. The smallest capillaries are 10–15 μm in diameter, formed by a single EC wrapped around a lumen through which only one or two erythrocytes can pass simultaneously. These capillaries are distributed throughout the vertebrate body with no cell being more than a few microns away from a capillary and even the poorest tissues having a few dozen cross-sections of capillaries in each square mm [26].

ECs of the cardiovasculature and lymphatic vasculature differ in several respects. Whereas ECs of blood vessels form tight junctions and adherens junctions that seal the entire edges of the cells to each other, the lymphatic ECs form focal adhesions along their borders with only occasional tight junc-tions. ECs of the two vasculatures are also different in their gene expression patterns and a number of protein markers (products of gene expression) have been identified that distinguish the two cell types [101].

As well as in the nature and morphologies of the ECs that form their structure, the vessels of the two vasculatures also differ in the structures that surround the single layer of endothelium. Blood vessels and capillaries are surrounded by a distinct proteinaceous basement membrane and muscle-like cells called pericytes (1). Lymphatic vessels are not surrounded by these fea-tures and instead their ECs are linked to fibres called "anchoring filaments" that extend well into the connective tissue that surrounds the vessels [36]. Larger blood vessels and also portions of larger lymphatic structures called procollectors are also surrounded by a muscle layer.

In this review we focus on the process of angiogenesis that results in the formation of blood-conducting capillaries of the cardiovascular system. Vascu-larization and the nutrient supply that it delivers is essential for tissue expan-sion as it has been demonstrated that a mass of cells cannot expand beyond about 1 mm in diameter in its absence [28, 63]. New capillaries are formed in response to specific protein signals released by growing tissues. These angio-genic signals cause capillaries to sprout from surrounding blood vessels and grow into the new tissue. New capillaries can also be initiated *de novo* by circulating stem cells from the bone marrow that are attracted to the tissue by the emitted angiogenenic signals [3].

3   What are the Key Events in Angiogenesis?

The key cellular activities in angiogenesis are cellular proliferation and mi-gration. Both these events are highly regulated and involve signals (generally proteins) that stimulate or inhibit the activities of cells or the proteins they

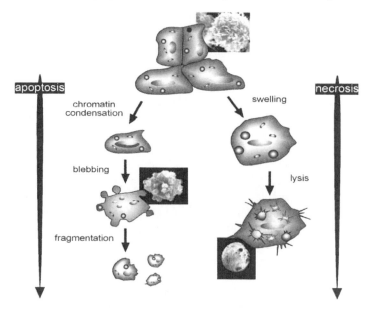

Fig. 2. Cell death. In apoptosis, or programmed cell death, shown on the left, the chromatin condenses and the cell blebs into smaller subunits. In necrosis, shown on the right, the cell first swells. The cell wall lyses and fragmentation results

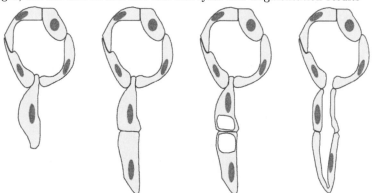

Fig. 3. Stages of angiogenesis. Capillary sprouting (cross section). From left to right, the invasion of the endothelial cell into the ECM, cell proliferation, onset of lumen formation, and, maturation of lumen

produce. Inhibition can be achieved by preventing proliferation or migration, but is also sometimes achieved by killing the cell, which dies by a process called apoptosis.

In apoptotic death the cell is fragmented into lipid membrane-enclosed fragments. The apoptotic fragments are engulfed by phagocytic cells like macrophages that roam the body and function as efficient garbage collectors. Apoptosis of ECs results in the regression of blood vessels, a process that is called pruning, which is not the same as necrosis. (Fig. 2).

New blood vessels can be formed by at least three different cellular modes of angiogenesis. First, and most discussed, is sprouting angiogenesis in which cells move away from the walls of an existing blood vessel and begin a new column of cells (a capillary) with a trajectory away from the original vessel and towards the angiogenic signal (Fig. 3). The second mode of angiogenesis occurs when circulating endothelial progenitor cells from the bone marrow, attracted by angiogenesis signals, move into the signaling tissue and form capillaries that link up with nearby existing blood vessels [95]. Also known as adult neoangiogenesis, this process resembles embryonic vasculogenesis during which the vascular system is initially formed.

A third mode of angiogenesis is a form of vascular remodeling, that involves the division of existing blood vessels into two by the process of intussusception. In intussusceptive angiogenesis ECs that line a blood vessel move inwards and meet in the lumen. The touching ECs then form junctions that eventually result in splitting the lumen into two. With time, other cells such as fibroblasts and pericytes move between the split vessel and thus two vessels are born from one [19].

The debate is not yet resolved as to how the capillary lumen is formed [21]. Some suggest that the lumen is an intracellular event in which vacuoles of contiguous ECs grow larger and finally fuse across cell borders to create a continuous lumen within the aligned endothelium of the capillary as is represented in Fig. 3 [31, 99]. This hypothesis is consistent with the observation that many capillaries are surrounded by a single EC. Others suggest that formation of the lumen is an intercellular event where certain cells die by apoptosis thus creating a lumen in their wake [69]. Yet another means of creating a lumen might be the extension of capillary walls around a lumen created by proteolytic degradation of the tissue material (extracellular matrix, abbreviated as ECM) into which the capillary is invading. These models are not mutually exclusive and could all operate together or in different tissue environments [31, 99].

From the molecular to the cellular level, each event and activity of angiogenesis is controlled by positive and negative regulators. Frequently the two regulatory modes are temporally shifted such that one mode initiates first followed by an increase in activity of the opposing mode. This general feature of normal physiological events limits the time period of the event and thus ensures that the system will again attain a steady state condition. The change that initiates angiogenesis can be an external perturbation such as wounding or irradiation or the development of a diseased state such as cancer or heart disease. These diseased cells also operate in the same tissue environment, and with largely the same molecular components, as normal cells. But, due to a change in one or more of their molecular components, the diseased cells' regulatory responses are altered such that a new steady state is approached. Because cells in the body interact and influence each others behavior, the new steady state that is achieved after disease is initiated includes changes in both diseased and normal cells of the body.

4 What are the Chemical and Cellular Contributions to Capillary Structure?

4.1 Endothelial Cells form the Capillary Wall

The EC is the key cellular component of the capillary that wraps itself around a central cavity called the lumen. There are many types of ECs that are recognized by their different patterns of gene expression. Cell edges overlap and are sealed by tight junctions, which are regions containing proteins that link the two opposing cell membranes and intracellular cytoskeletal structures in each cell. By this structural integration the endothelium seals the tissues from direct contact with the blood and its contents. Gases pass rapidly from the blood through the ECs to the tissues beyond. The ECs have at least two adaptations to promote movement of nutrients from the blood to the tissues. First, the cells carry out an active process of pinocytosis, which refers to the pinching off into the cell of small vesicles from the cell surface membrane that contain blood fluids. These pinocytic vesicles move from the lumenal (blood side) surface to the ablumenal (tissue side) surface of the endothelium and the vesicle contents are released into the tissue.

To increase the speed of delivery of the nutrient-containing blood fluids into the tissues the EC has a unique feature referred to as fenestrae ("windows") in which the lumenal and ablumenal membranes have fused to create pores of about 150–175 nm in diameter [10]. The extent of fenetration and the size of the fenestrae in a blood vessel varies with the host tissue and with the presence a variety of drugs and hormones. Capillaries that are not fenestrated are called continuous capillaries. Fenestrated capillaries are divided into two types depending on whether the basement membrane surrounding the capillary is present (fenestrated capillary) or absent (discontinuous capillary). Discontinuous capillaries form in tissues such as the liver where exchange of materials between blood and tissue is rapid.

ECs are also the means by which white blood cells and endothelial progenitor cells from the bone marrow recognize a region of the body in which changes are occurring – e.g. disease or inflammation. In response to signals from the changing tissue, nearby ECs expose certain receptors on their surfaces to which the blood cells attach and then move through the EC layer into the tissue. The movement of white blood cells (leukocytes) has been well studied and involves a process called diapedesis, which literally means "walking through". Recently, a "transmigratory cup" structure has been observed in ECs that is believed to be the means by which leukocytes are directed through the EC cytoplasm [13]. Although it is yet been demonstrated, endothelial progenitor cells may also pass through the endothelium by way of the transmigratory cup.

4.2 Pericytes Surround the Endothelial Cells and Provide Structural and Functional Support

Pericytes surround the ECs in capillaries. Apposition of these two cell types is intimate as evidenced by their interdigitated cell surfaces [9]. Pericytes, sometimes called Cells of Rouget, are of mesechymal derivation and are believed recruited by the ECs. The ECs promote the differentiation of these undifferentiated mesenchymal cells into pericytes [43]. The traditional view is that the ECs first form the nascent capillary and then recruit pericytes to surround them [51]. However, more recent evidence suggests that the pericyte also partners with the EC in forming the capillary channel and lumen. For example pericytes were found near the tips of capillaries in newborn mouse retina and in tumors there are regions of capillaries that contain only pericytes [86].

Pericytes are believed to play two important roles in capillaries. The first, based on their observed location in tissues, is structural. This conclusion is based on observations of location. For example, there are few pericytes located around capillaries in the muscle where there are many other cells that can provide support for the fragile ECs that form the capillary. By contrast, there are many pericytes around capillaries in the brain and the feet and distal legs where it is postulated more mechanical support is needed to maintain lumen structure. Pericytes are also located in identifiable positions in capillaries, such as at the junctions of endothelial venules and over the gaps between ECs that are created during inflammation (reviewed in [105]). The presence of the pericytes stabilizes the vessel wall. Capillaries surrounded by pericytes are much less likely to regress than capillaries without these cells.

The second role of the pericyte is communication with the EC that results in a coordinated course of capillary development. The communication is mutual and involves each cell type either inhibiting or promoting proliferation of the other. For example, the pericyte secretes inhibitors of EC growth that would have the effect of suppressing the lateral expansion of an already formed capillary [105]. During periods of capillary growth, such as when oxygen levels are low, the pericytes secrete vascular endothelial growth factor (VEGF) that stimulates EC growth [123] and angiopoietin-1, a survival factor for ECs [23]. Conversely, the EC secretes a growth factor called platelet-derived growth factor (PDGF) that stimulates pericyte growth [40]. VEGF secreted by pericytes also stimulates pericyte cell growth, a phenomenon known as autocrine regulation of cell growth. The impact of this close relation between ECs and pericytes is evidenced by the observation that treatment with an inhibitor of EC growth was unsuccessful in causing regression of tumor capillaries and resulted in an increased production by pericytes of angiopoietin, a survival factor for ECs. By contrast, treatment with an inhibitor of both endothelial and pericyte growth resulted in capillary regression [23].

4.3 The Basal Lamina Encases
the Endothelial Cells and Pericytes

Capillaries are surrounded by a proteinaceous membrane called the basement membrane or basal lamina [79], and discussed in [5]. The major protein and proteoglycan components of this membrane include the proteins laminin, type IV collagen, entactin/nidogen, and fibronectin, and a heparan sulfate proteoglycan called perlecan. These components are produced by the ECs and surrounding pericytes and are organized in the membrane so as to give it the characteristic lamina structure observed when tissue sections are analyzed by transmission microscopy. The basement membrane is probably formed by the cells during the process of angiogenesis and is found to surround new capillaries up to, but not including, the growing tip [52].

5 Interaction of Endothelial Cells
with Their Environment

5.1 The Integrins and the Extracellular Matrix

The tip of the capillary is the location of most of the cell proliferation and movement in the forming capillary [12]. Unlike the cells in the column behind the tip, which are in contact with the basement membrane and frequently associated with pericytes, these cells are in contact with the ECM of the tissue into which the capillary is growing. The ECM consists mainly of proteins and proteoglycans and is the source of many cues for the invading ECs and pericytes. These cues are received by cell surface receptors that interact physically with the proteins in the matrix. The cells integrate the information gained from a variety of cell surface receptors to "sense" their environment.

The composition of the ECM varies between tissues. However, the major protein components of most extracellular matrices are the collagens, laminins and fibronectin. These proteins are recognized by a class of receptors called the integrins [96]. The integrins are a family of related proteins situated in the cell surface with their longer axis at 90° to the plane of the membrane. As for all proteins, the integrins are polymers of amino acids that fold into several defined and distinguishable structures referred to as domains. Being transmembrane proteins, the integrins have three major domains of structure identified as the extracellular, transmembrane and intracellular domains. The extracellular domain interacts with the ECM protein and the intracellular domain creates signals inside the cell in response to this extracellular interaction.

Functionally, each unique integrin receptor consists of a different combination of one α and one β subunit. The subunits are encoded by different genes. So far, 18 different α subunits and 4 different β subunits have been identified in mammals. Together, the α and β subunits forming the active receptor that can transmit information bidirectionally across the membrane [96, 121]. Each combination of α and β subunit provides a different specificity of binding to one or more ECM proteins and a different specificity for interacting with intracellular signaling molecules. Although all possible combinations have not been identified, at least twenty-four different $\alpha - \beta$ integrin combinations have been characterized, subsets of which are on the surfaces of every cell in the body. Aptly named, the integrins function to integrate the cell's behavior with its environment.

ECs create a large part of their environment. Thus, they produce and secrete many proteins into the ECM and basement membrane that surrounds them. When removed from their normal tissue environment and placed in culture, the ECs synthesize and secrete ECM proteins. These proteins adsorb to the plastic dishes in which the cells are cultured and the cells attach themselves to the ECM proteins as they were attached in vivo. Much has been learned about the interaction of cells with ECM proteins from studies of cells in culture. For example, the ECM promotes EC migration, proliferation, survival, and morphogenesis and tubes can be formed in culture by ECs in the presence of the appropriate combination of ECM proteins [18, 88]. As is expected from the observation that cells express a defined number of integrins on their surfaces, the cellular response to ECM proteins is saturable. However, for some cellular functions, such as migration, the EC response is biphasic with optimal activity in a middle-range of ECM protein concentrations [18, 88].

Once secreted, the ECM proteins interact in defined ways to form a larger three-dimensional assembly. Although each $\alpha - \beta$ heterodimeric integrin molecule recognizes only a portion of its ECM protein ligand, the combination of integrins and other receptors on the cell surface provide the EC with a means of gauging the larger structural features of the ECM assembly [48, 102, 107]. This recognition might be the result of the combination of receptors on the cell surface that are in contact with their ligands. However, the flexibility of proteins also plays a role. When ECM proteins interact in macromolecular assemblies new epitopes are exposed for the EC to recognize. These new epitopes can result from the close apposition of polypeptide chains from two different proteins, but are also likely the result of local changes in the structure of particular proteins promoted by their interaction with other proteins in the macromolecular assemblage of the ECM [106] and references therein.

6 What are the Extracellular Events Leading to Vascularization?

The interaction between EC and ECM is bidirectional. The ECM influences the morphology and function of the EC and the EC create and remodel the ECM. Essential to the remodeling are proteases secreted by the EC that cleave ECM protein bonds and result in the eventual destruction (decay) of the proteins. In some cases, the action of proteases may expose new sites on

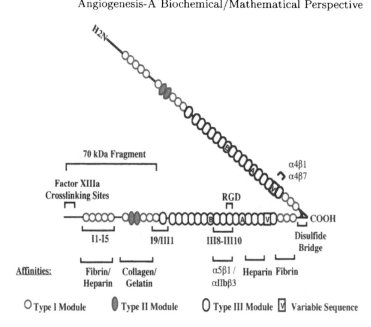

Fig. 4. Structure of fibronectin. Each symbol (oval, circle or square) represents a sequence of amino acids that are identified as a structural entity or domain. The RGD sequence that interacts with the integrins is identified in the bottom molecule (From Magnusson, and Mosher, 1998 Arteriosclerosis, Thrombosis, and Vascular Biology 18 1363–1370, permission pending)

the ECM for cell interaction. Proteases can also release active polypeptide fragments from the ECM proteins. For example, endostatin, an inhibitor of angiogenesis, is a C-terminal (-carboxyl or -COOH end) fragment of collagen XVIII released by the action of the protease plasmin [47].

Other proteins that have become associated with the ECM after its assemblage can be released by protease action. These proteins are also originally secreted by the EC or surrounding cells such as pericytes. They are growth factors[1] growth inhibitors, survival factors and morphogenetic factors for ECs. Thus, the constant remodeling of the ECM in vivo is achieved by a continuous interaction between the cell and its environment that allows the EC to pick up cues previously laid down by itself and by its EC neighbors (autocrine cues),

---

[1]The term "factor" refers to the historical means by which these extracellular regulatory proteins were first identified as components in mixtures of proteins such as serum or tissues extracts. When first identified, the activity, such as stimulation of angiogenesis or growth, was referred to as being caused by a factor present in these protein mixes. The protein, thus identified as a growth factor, angiogenesis factor, etc. was later purified and the protein sequence and its gene identified precisely. However, in many cases the designation of factor has remained associated with these regulatory proteins.

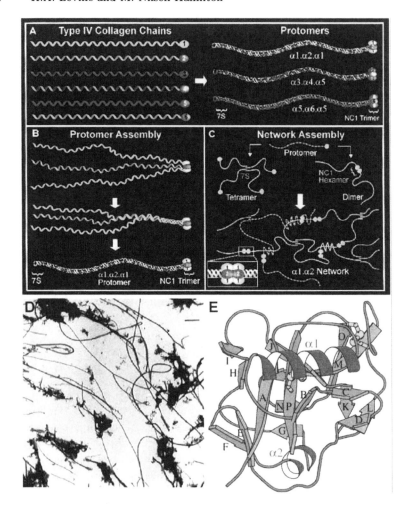

Fig. 5. Assembly and structure of collagen. Type I collagen chains (A) form triple helices called protomers (B). Protomers assemble into large macromolecular complexes in the ECM (D) as also seen in electron micrographs of extracellular matrix protein preparations. Proteases cleave endostatin from the C-terminus of the assembled monomers. The molecular structure of an endostatin molecule is shown in which the polypeptide chain is represented as a ribbon diagram (E). (The blue arrows refer to the β sheet secondary structure while the orange tubes correspond to the α helix secondary structure of the protein). (From Sundaramoorthy etal 2002 JBC 277 31142-53, Bätge et al. 1997 J Biochem 122 109–115 Hohenester et al. 1998 EMBO J 17 1656–64, permission pending)

cues from neighboring unlike cells such as pericytes (paracrine cues) and cues delivered by the blood stream from distant cells and tissues (endocrine cues).

6.1 Individual Matrix Proteins

Amongst the myriad of ECM proteins, several stand out as having a major impact on angiogenesis. One such protein is fibronectin (FN; Fig. 4). This protein was first named LETs (large external transformation-sensitive protein) when it was found to be lost when cells in culture become transformed (cancer-like) [97]. Fibronectin is synthesized by almost every cell type in the body and becomes a component of the ECM laid down when cells are taken from the body and cultured in the laboratory. The cells interact with fibronectin through a variety of integrin heterodimers including $\alpha 3 : \beta 1$, $\alpha 4 : \beta 1$, $\alpha 5 : \beta 1$, $\alpha 8 : \beta 1$, $\alpha V : \beta 1$, $\alpha V : \beta 3$, $\alpha V : \beta 5$, $\alpha V : \beta 6$, $\alpha 4 : \beta 7$, and $\alpha V : \beta 8$ [52]. In each of these interactions cells recognize the very small sequence domain on fibronectin that is typified by the three amino acid sequence RGD (Arg-Gly-Asp) [35]. The importance of fibronectin to angiogenesis is evidenced by the observation that mice with the fibronectin gene inactivated die in utero with deformed vasculature [34]. Further analysis of the process of vasculogenesis and angiogenesis in these knockout mice revealed that the presence of fibronectin is critical for correct morphogenesis of the heart and blood vessels, but not for the initial differentiation of stem cells and conversion of progenitor cells to become endothelial cells [119]. Although these observations were of vasculogenesis rather than angiogenesis, which occurs in the adult, it is likely that in angiogenesis fibronectin is also required for formation/stabilization of the vessel lumen and may not be necessary for the early events in angiogenesis such as cell migration, differentiation and tube formation [106].

The collagens have long been recognized as structural proteins of tissues and cartilage (Fig. 5). They are long molecules that associate as trimers with a helical central structure and nonhelical ends. These assembled collagen structures form a $\sim$1.5 nm diameter by 300 mm-long rod resembling a thread that is unraveling at each end. Their thread-like structures allow them to contribute to microscopic collagen fibres containing many collagen trimers that are stacked together in a staggered configuration to form a collagen bundle that becomes the basis of cartilage and other structural features of the body. The strength of the collagen bundles is augmented by the individual collagen molecules being chemically cross-linked to one another during formation of the fibres and further stabilized by the association of other proteins such as decorin [34]. Collagens are an important structural component of the basement membrane that lines the capillaries and blood vessels and is formed during angiogenesis. Mutant mice that lack collagen I die in utero with evidence of rupture of their major blood vessels [66]. Mice with mutations in collagen III also die young with ruptured blood vessels [65]. Mutations in collagen III result in disordered collagen I fibrils and it is believed that collagen III is required for the formation of structurally sound collagen I fibrils [38].

The importance of the collagen I fibrils to angiogenesis is evident because agents that inhibit collagen crosslinking also inhibit angiogenesis [49]. From these observations the basement membrane has been identified as a possible target for controlling tumor growth by inhibiting angiogenesis [67].

As well as being an important component of the basement membrane that surrounds the growing capillary, collagens are part of the ECM into which the growing capillaries move. Cell migration is associated with the release of proteases that cleave proteins in the ECM to allow the cells to enter this space. Protein cleavage alters the exposed epitopes and, for collagen IV, cleavage by certain proteases results in the exposure of a cue, called a cryptic migratory site, that promotes endothelial cell migration in the direction of the cleaved collagen, a process known as haptotaxis [39, 122].

Laminins are important components of the extracellular matrix for angiogenesis and many other events in tissue morphogenesis. These proteins are heterotrimers made of one of each of three different types of subunits named alpha, beta and gamma. A large proportion of the length of the laminin heterotrimer is a coiled coil that forms a fibrillar structure. In all, 15 different laminin heterotrimers have been identified that consist of different combinations of six α, four β and two γ subunits. Laminin 8 ($\alpha 4 - \beta 1 - \gamma 1$) predominates in the basement membrane of capillaries. Mice that do not contain an active laminin $\alpha 4$ subunit gene show impaired microvessel development [113]. The further polymerization of laminins into the larger structures found in the ECM is believed to be promoted by their calcium binding N-terminal (LN) domains [71]. Laminin also interacts with other components of the basement membrane and is essential for the assembly of the macromolecular complex ECM. Basement membrane does not assemble in the absence of the $\gamma 1$ subunit of laminin despite the presence of other basement membrane components such as type IV collagen, nidogen and perlecan [62]. In addition to laminin at least two other extracellular calcium-binding proteins play important roles in angiogenesis as part of the ECM. These are fibrillin-1 and fibulin-1. Mice containing mutations in each gene die around birth with hemorrhages of many blood vessels [38, 56, 91].

7 Soluble Proteins that Modify the ECM and Influence EC Function: Proteases and Protease Inhibitors

The condition under which the physiological activities of cells and tissues are maintained in equilibrium is referred to as homeostasis. To achieve homeostasis cells receive and respond to extracellular cues in the form of molecules that move in their immediate environment. These extracellular signals guide cells to decisions regarding the rate of their metabolism, whether they proliferate, remain quiescent, or undergo apoptosis, what genes they activate or deactivate, how they distribute proteins on their surfaces and throughout the cell, which cellular proteins are active, what shape they adopt and what proteins they secrete. Receptors (also proteins), most of which are located on the cell surface are the means by which cells recognize extracellular cues. Like the integrin receptors, most receptors are transmembrane proteins with extracellular, transmembrane and intracellular domains. Some receptors are located entirely inside the cell to recognize hydrophobic cues that move readily through the lipid membrane of the cell surface. Each cell's ability to respond to the cues in its environment depends in part on the receptors that it produces and places appropriately to receive external cues. The other necessary component for each cellular response is the presence and correct intracellular placement of the components of the signal transduction pathways that transmit the extracellular signal to activate an intracellular event.

The ECM is a dynamic structure. It is actively maintained and, when necessary, remodeled by the cells imbedded in and around it [29]. Cells regulate the content of the ECM by secreting new ECM protein, and inhibitors of these proteases. ECM proteins are degraded by proteases (also called proteolytic enzymes or proteinases) that cleave the polypeptide chains that constitute proteins, thereby creating smaller polypeptides. The site between two amino acids on a particular protein that is cleaved by a protease is determined by the specificity of that protease for the amino acid sequence around the cleavage site (sissile bond) and by the availability of that site to the protease. Active proteases are sensitive to environmental factors such as pH. Thus, although in the active form, the protease might only perform its activity in specific locations in the ECM that possess the appropriate conditions for optimal protease activity.

Different proteases have different specificities. For most proteases there is degeneracy in the amino acid sequence recognized for cleavage. There may be several or even many sites on a protein recognized by a particular protease. Cleavage(s) by a protease to release two or more polypeptides can reveal other sites for cleavage by the same or by another protease. Cells secrete many different proteases with different specificities with the result that ECM proteins can eventually be degraded to their amino acid constituents or to small polypeptides that are taken up by the cells for complete degradation. The resulting amino acids can be used by the cells for synthesis of other proteins.

Although degradation of ECM proteins eventually goes to completion resulting in "recycled" amino acid and peptide products that provide nutrients for the cells, some cleaved fragments are used by the cells to maintain homeostasis and as cues to signal changes in the ECM. Two examples are endostatin, which is a cleavage product of collagen, and angiostatin, which is a cleavage product of plasminogen. In the latter case, cleavage of plasminogen by plasminogen activator results in two functional products, which are plasmin and angiostatin. Plasmin is a potent protease that degrades the ECM proteins. Degradation of ECM proteins releases many soluble factors that had been previously deposited in the ECM.

In their initial forms, when secreted by cells, most proteases are in an inactive condition referred to as the proform. Conversion of the proform to the catalytically active form of the protease usually involves cleavage of the proform to release a terminal fragment. This cleavage can be achieved in several ways. The proforms of some proteases have very low catalytic activities that can sometimes also be activated by specific extracellular conditions, such as particular pH ranges, to self-cleave. This is referred to as autocatalysis. For some proteases proform activation is achieved upon cleavage by one or more other types of proteases. Many examples exist of cascades of protease activation where the activation of one protease results in the cleavage and activation of another of a different type. The activation of plasmin by plasminogen activator is part of such a proteolytic cascade [64]. This cascade is also regulated by a positive feedback mechanism, in which plasmin activates plasminogen activator, that results in an exponential explosion of plasmin activity initiated by a small amount of catalytically active plasminogen activator.

When plasminogen is cleaved to form plasmin, the N-terminal fragment released by the action of plasminogen activator, called angiostatin, is an inhibitor of angiogenesis. Thus, by the single action of secreting the protease plasminogen activator, cells cause the activation of plasmin, degradation of the ECM, release of a number of growth and angiogenesis factors from the ECM and release of angiostatin an inhibitor of angiogenesis. The consequence of this complex response to plasminogen activator release is a temporary deviation from homeostasis. For example, angiogenesis is stimulated by an increase in active plasmin and the growth factors released from the ECM. Certain other proteases also release angiostatin from plasminogen [90].

Synthesis and release of proteases is highly regulated temporally such that, after a perturbation that results in the increased expression and release of proteases, the production of these proteases soon decreases to the low original basal level(s). Without continued release of plasminogen activator and other proteases, homeostasis is soon reestablished by the activity of angiostatin and other inhibitors of angiogenesis.

In addition to the cells tightly controlling the rate at which proteases and their proforms are synthesized and secreted, proteases are controlled by protease inhibitors that are secreted by EC and other cells. Examples of inhibitors relevant to angiogenesis are the plasminogen activator inhibitors (PAI-1 and PAI-2), tissue inhibitor of metalloproteinases (TIMPs -1 through 4). The balance of protease and protease inhibitor secreted by the population of cells in the tissue is critical to maintaining tissue structure and function. Too little protease activity prevents the cells from remodeling the ECM, for example to allow the EC to migrate in angiogenesis. Too much protease activity results in disintegration of the tissue. Consequently the synthesis and secretion of protease inhibitors is also tightly controlled by the cells in response to many cues such as growth factors and growth inhibitors. In some cases the cells integrate signals from several regulatory factors to establish a rate of production and secretion of proteases and their inhibitors [111, 112].

The role of proteases in angiogenesis is more complex than their catalytic action on ECM proteins. When bound directly to the cell surface they are also involved in regulating cell movement and other cell responses. For example, the urokinase-type plasminogen activator receptor (uPAR) is a specific receptor for urokinase plasminogen activator (uPA) that is linked to the cell surface by a glycosyl phospholipid tether. This receptor, that lacks an intracellular domain interacts with other cell surface receptors with intracellular domains, such as the integrins, and the epidermal growth factor receptor, and thereby regulates cellular activities that include proliferation, cell shape and cell migration [84]. uPAR is localized to the leading edge of the cell surface of migrating cells [24]. Plasmin also binds to several molecules on the cell surface, including a histidine-rich glycoprotein, annexin-II, gangliosides and $\alpha V \beta 3$ integrins and promotes cell migration by a mechanism that requires it to be catalytically active [54, 109]. The close association between plasminogen and uPA on the cell surface increases the probability that plasmin will be activated and provides the cell with a leading cutting edge for penetrating the ECM. Interestingly, angiostatin, the portion of plasminogen that is cleaved off by uPA to produce active plasmin, also binds to $\alpha V \beta 3$ integrins and inhibits the cell migration promoted by plasmin [109].

8 Growth and Angiogenesis Factors: Factors that Stimulate Angiogenesis

Angiogenesis is regulated by growth factors and angiogenesis factors, which are proteins that stimulate cellular functions by binding to and activating specific cell surface receptors (Fig. 8). Some of these proteins, such as VEGF and angiopoietin, act specifically on ECs. Other proteins, such as FGF, angiogenin, EGF, PDGF, and CXC cytokines with ELR motifs also stimulate proliferation of other cell types in the body including those cells that contribute to new tissue formed during repair. Similarly, there are many protein inhibitors of angiogenesis, some of which seem specific for ECs (endostatin) and others that also affect the behavior of other cells (angiostatin, PEDF, TGFα, TNF, angiopoietin 2, CXC cytokines without ELR motifs).

In most cases, growth factors and angiogenesis factors are produced locally. Their production or release is regulated by changes in the tissue environment that characterize conditions requiring angiogenesis. These changes occur when a tissue is wounded or damaged resulting in the need for vascularization of the new tissue produced to repair the damage. Events that regulate the production of growth factors include hypoxia (decreased oxygen available to the damaged tissue), breakage of cells in a wounded tissue, released proteases, and entry of cells of the adaptive immune response that release cytokines. Some of these events (hypoxia, cytokines) initiate changes in gene expression to produce more angiogenesis factors. Other events result in the release of angiogenesis factors from the ECM (proteases) or the cells (cell breakage). Both types of events are important for regulating angiogenesis.

Many angiogenesis factors have been identified. They include proteins that signal changes in behavior of ECs or other cells that regulate angiogenesis. A balance of positive and negative signals for cellular behavior is a hallmark of biological control mechanisms that moderates the extent of the cellular response and ensures a limited time of response. Of all the angiogenesis factors, a central player is vascular endothelial growth factor (VEGF), which provides a positive signal for angiogenesis by promoting EC proliferation and migration towards a region of higher VEGF concentration, a process known as chemotaxis [37, 80]. VEGF also promotes EC survival under adverse conditions, such as lack of nutrients or other growth factors, and it promotes tube formation by ECs. VEGF production is increased in cells under hypoxic conditions. Although encoded by a single gene, there are several forms of VEGF (called isoforms) that vary in the length of their primary (polypeptide) sequence and that have different propensities for interacting with the ECM due their secondary and tertiary (folded) structures.

Fibroblast growth factor-2 is also a potent angiogenesis factor. Also called basic FGF (bFGF), FGF-2 is a member of a large family of related proteins of which there are at least twenty-four members encoded by different genes. FGF-2 is an unusual extracellular protein because it does not have the typical N-terminal (amino, NH_3 sequence (signal sequence) required for secretion by the conventional secretory pathway that involves the endoplasmic reticulum and the golgi apparatus. Instead FGF-2 is released in vesicles shed from the cell surface [110]. FGF-2 shedding is stimulated by serum, that is produced on wounding as a result of blood clotting.

As well as acting independently, FGF-2 and VEGF can act together to stimulate angiogenesis by several means. For example, FGF can induce the increased expression and production of VEGF by endothelial cells [103]. Some isoforms of VEGF can displace FGF-2 from the ECM with the resulting effects on cell proliferation being directly stimulated by the freed FGF-2 rather than by VEGF [53]. When present together VEGF and FGF-2 act synergistically to stimulate angiogenesis [4].

FGF-2 and the 165 kDa isoform of VEGF bind heparan sulphate proteoglycans (HSPGs) that are found on cell surfaces, in the ECM and in body fluids. Some HSPGs are located on the cell surface where they can promote VEGF and FGF-2 actions [98]. Syndecan and glypican-1 are two well-described cell surface HSPGs that interact with FGF-2 and VEGF to promote the efficiency of activation of their respective signaling receptors [14, 50, 94, 114]. Perlecan, an HSPG located in the ECM, has both positive and negative effects on bFGF signaling. But, removal of the heparan sulfate component of this proteoglycan results in impaired wound healing and angiogenesis and in diminished FGF-2-induced tumor growth in transgenic mice [125]. These results suggest that perlecan also promotes FGF-2 activation of angiogenesis.

Some HSPGs inhibit angiogenesis. For example, heparan sulphate proteoglycans in the aqueous humor of the eye bind FGF and VEGF and prevent these angiogenesis factors from binding their receptors on the cell surface and activating the cellular events that lead to angiogenesis [25]. Similar reservoirs of growth factors are believed to be bound by HSPGs in the extracellular matrix [116]. Perturbations that release growth factors from HSPG-bound reservoirs change the balance of signals to ECs and can initiate angiogenesis. Another inhibitory effect of HSPGs comes from type VIII collagen, a hybrid collagen/HSPG that is located in the basement membrane and is the source of the angiogenesis inhibitor, endostatin.

9 Inhibitors of Angiogenesis

At least two types of angiogenesis inhibitors are produced as a result of proteolysis of ECM proteins. These are the endostatins, which are released from types VIII and XV collagens (Fig. 5), and angiostatin, which is released from plasminogen. Both angiostatin and the endostatin derived from type VIII collagen bind to HSPGs and are likely also trapped in the ECM to be released secondarily upon degradation of the HSPGs that hold them. Different proteases are responsible for creating these inhibitors, with endostatins cleaved from the collagens by cathepsin L or matrix metalloproteases (MMPs) and angiostatin cleaved from plasminogen by plasminogen activator. These proteases are produced in response to tissue damage and their expression is stimulated by FGF-2 [27, 82, 93].

The inhibitors released by protease action bind a variety of proteins and HSPGs in the ECM and on cell surfaces. Angiostatin binds several proteins on the cell surface, including angiomotin, the subunits of cell-surface ATP synthase, annexin II and the $\alpha V\beta 3$ integrins. By interacting with these cell surface proteins, angiostatin may inhibit angiogenesis by inhibiting EC migration [57, 73, 109, 115]. Endostatins also bind specific receptors on the cell surface. The two endostatins (-V and -VIII) are similar in three dimensional structure but only 61% identical in primary sequence, which results in different combinations of amino acid side-chains being exposed on their surfaces [100]. Consequently, these molecules demonstrate differences in affinities for molecular targets on the cell surface and in the ECM.

The best studied endostatin is endostatin-VIII that binds the $\alpha 5\beta 1$ integrin receptor through which it inhibits adhesion to the ECM, causes disassembly of the focal adhesions that hold the cell to its substratum and decreases the secretion of ECM proteins by ECs [120]. These cellular responses are regulated by the integration of a multitude of intracellular signaling events [2]. Other cell surface molecules such as glypican (an HSPG), KDR (the VEGF receptor) and the TNF receptor may also be involved in regulating the cellular response to endostatin [8].

An obvious means of inhibiting angiogenesis is to inhibit the proteases that promote cell migration and proliferation. As expected, protease inhibitors of angiogenesis include inhibitors of plasminogen activator (PAI-1 and PAI-2) and of MMPs (TIMPs). However, recent studies revealed an unusual twist when it was found that, rather than mediating its effect on angiogenesis by inhibiting the activity of MMPs, TIMP-2 acts directly on the EC by binding $\alpha 3\beta 1$ integrins to activate phosphatases that remove phosphates from the intracellular domains of the FGF and VEGF receptors [104]. Dephosphorylation inactivates these receptors and results in decreased levels of intracellular signals that promote angiogenesis.

Inhibition of angiogenesis also occurs by pure competitive mechanism, whereby an inhibitor binds an angiogenesis factor to prevent it from binding to its cell surface receptor by which it stimulates angiogenesis. Some ECM proteins, other secreted proteins and HSPGs fit in this category. However, their roles are often quite broad in that they bind many growth factors and influence many cellular events. By contrast, sFLT is a very specific competitive inhibitor of angiogenesis. This protein is a product of the same gene as the VEGF receptor, FLT-1. However, the alternative mRNA transcript that encodes sFLT-1 is shorter than the mRNA that encodes FLT-1. As a result, synthesis of sFLT-1 terminates before the transmembrane sequence of the full-length receptor and the resulting sFLT-1 is secreted by ECs as a soluble extracellular protein. This secreted extracellular domain of the VEGF receptor binds to VEGF and thus sFLT-1 competes with the cell surface FLT-1 receptor for VEGF. sFLT-1 expression is regulated differently from the expression of FLT-1 and thus, it is likely that EC regulate their ability to respond to VEGF in part by secreting sFLT-1 [70].

Other cells such as haematopoietic cells also produce inhibitors of angiogenesis. Growth factors and other cellular regulators produced by hematopoietic cells are collectively referred to as cytokines. IL-12 is a cytokine produced by dendritic cells, macrophages and monocytes that inhibits angiogenesis in vivo [117]. The mechanism of this inhibition appears to be indirect and involves other cytokines and matrix metalloproteases. IL-12 stimulates the secretion of IFN by T lymphocytes and natural killer (NK) cells. IFN stimulates the production of the chemokines CXCL9 and CXCL10 by CD4+ lymphocytes. These chemokines suppress the production of MMP9 by endothelial cells and thereby inhibit angiogenesis [72].

10 Cellular Events that Characterize Angiogenesis

10.1 Proliferation

Most cells in the human body are quiescent, which means that they are not proliferating. Proliferation of EC and other cells is stimulated by growth factors. The ability of a cell to respond to a particular growth factor is determined by the presence of specific receptors on that cell's surface. Growth factor receptors transmit a signal from the outside of the cell to the inside that results in changes in the expression of genes that control cell proliferation. Different cell types are identified by the growth factor receptors that they express on their surfaces. ECs present FGF and VEGF receptors and therefore proliferate in response to these two growth factors.

Proliferating cells pass through defined phases of cellular activity before they divide to form two cells (Fig. 6). The phases of the cell cycle are characterized by the genes that are expressed and the protein activities that are present in the cell during that period. These phases are referred to as G_1 (gap 1), S (DNA synthesis), G_2 (gap 2) and M (mitosis). Growth factors bind to the extracellular domains of their specific receptors and initiate cascades of intracellular signals that target particular genes to initiate the growth cycle. Quiescent cells are viewed as residing in a fifth growth phase referred to as G_0. Whereas, with proper nutrition and other requirements, a cell can remain in G_0 for an indefinite period, the cell cycle (G_1, S, G_2 and M) takes between 12 and 70 hours to complete. The variability in cell cycle length probably partly depends on the cell type. However, even for a single cell type there is some variability in cycle time that occurs at specific periods in the cell cycle. For example, directly after cell division (mitosis) is a period in G_1 referred to as G_1-pm that takes 3–4 hours in cells studied so far (see [22] and references therein). Passage through this phase is highly dependent on the presence of growth factors. If growth factor receptors are not activated during this period the cell diverges from the growth cycle and enters G_0. The presence of growth factors allows the cell to pass through a restriction point (R) to the second part of G_1 for which transit does not depend on the presence of growth

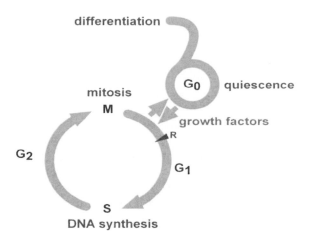

Fig. 6. The cell cycle. Cells proliferate by progressing through G1, S, G2 and M. Current thinking is that cell differentiation is preceded by cell entry into the G_0 or quiescent state

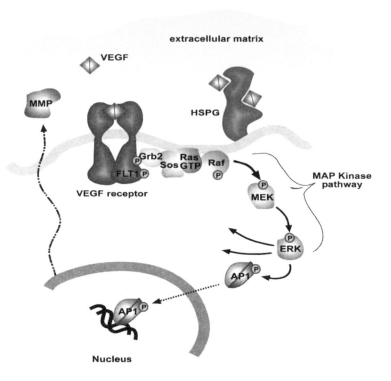

Fig. 7. VEGF signalling. Here a molecule of VEGF is shown bound to its dimeric receptor (*blue*). Within the cell cytoplasim a signal transduction cascade is shown resulting in the activation of a protease (MMP). The transcription factor (AP1) which enters the nucleus to activate transcription of the MMP gene. This results in the cellular expression of the protease

factors. This portion of G_1, called G_1-ps is highly variable in its length with some cells spending up to 20h before reaching the next phase of the cell cycle, the S phase.

Proteins called cyclins are central regulators of the cell cycle and the genes encoding certain cyclins are primary targets of growth factor-initiated signal transduction [74]. The cyclins are regulators of ser/thr protein kinases, enzymes that use ATP to add phosphate to serine and threonine residues on specific proteins that effect transit through the cell cycle. Phosphorylation is a frequent means of controlling enzyme activity and protein function. Addition of one or more negative charges due to the addition of phosphate(s) to strategic location(s) on the protein molecule alters the local electrostatic configuration and the structure of the phosphorylated protein with the result that the protein's activity changes. Thus, growth factors increase expression of the cyclin D1 gene[2], which is followed by increased production of the cyclin D1 protein by the process of translation. Cyclin D1, in turn, activates the protein kinases Cdk4 and Cdk6. Cyclin D1 is rapidly degraded during the subsequent S phase. In a similar manner cyclins A and B control the transit though S, G_2 and M, each synthesized at the appropriate point in the cell cycle and rapidly degraded prior to or during the next stage.

10.2 Apoptosis

EC death is tightly controlled by environmental signals including cytokines and growth factors. A cell contains many components that could be either toxic to the cells surrounding it or could activate an inflammatory response in the tissue. To avoid the release of intracellular material, cells die naturally by a mechanism called apoptosis. Apoptotic death is orchestrated by a regulated sequence of cellular events that involve a cascade of intracellular proteases called the caspases. Apoptosis can be initiated in cells by specific cytokines that activate receptors, which initiate the caspase cascade, or by stress caused by events such as oxidation of surface or intracellular proteins or other molecules (oxidative stress) that results in activation of caspases via cytochrome c release from the mitochondria. Unlike necrotic cell death, during which the cells lyse and release their contents into the surroundings, apoptotic cell death involves the cells breaking into smaller portions that are surrounded by a cellular membrane and that can be engulfed by the circulating white blood cells. The absence of growth factors results in apoptosis of ECs and most other mammalian cells. Although the mechanism of this regulation is not clearly defined, it is suggested to be mediated by the release of reactive oxidative molecules by growth factor-starved cells [89]. Apoptosis of ECs is also inhibited by shear stress by a mechanism that is ill-defined but is reported to involve the MAP kinase signal transduction pathway (a series of sequentially acting protein kinases) and an inhibitor of caspases [92, 108].

10.3 Migration

Some growth factors, such as VEGF and FGF, also stimulate cells to migrate. The direction of migration is up the growth factor gradient (chemotaxis) if one exists. Migration is also regulated by signals (signal transduction) emanating from the growth factor receptor. In this case, the signals result in the

---

[2]To increase expression of a gene means that the gene is activated and more transcript is produced. The transcript becomes messenger RNA (mRNA) that is then translated to protein. Frequently the term "increased gene expression" is used more generally and refers to increased mRNA encoded by a particular gene. As the steady level of a particular mRNA depends on the rate of its synthesis and degradation and both of these are controlled events, the reader can not be confident that a reference to increased gene expression truly reflects increased transcription from that gene unless experimental evidence is presented to verify this conclusion.

modification and consequent activation of proteins that regulate cell shape and cell adhesion to the ECM. These proteins constitute the cell cytoskeleton, a diverse group of proteins that form large multiprotein complexes. Some of these complexes (such as formed by actin and tubulin) are long fibers that can extend the entire length of the cell and that grow by the addition of more protein subunits to one end. Others form large multiprotein complexes that organize at specific sites on the membrane and form connections with the integrins and thus also with the ECM proteins bound by the integrins. These complexes are the molecular basis of the focal adhesions. The presence of a growth factor at only one side of the cell results in localized activation of growth factor receptors, which in turn locally activates the growth of actin fibers and assembly of focal adhesions. Local growth of actin fibers results in extension of the cell membrane towards the growth factor to form a cellular structure called a pseudopodium. Focal adhesions are formed at the tips of the pseudopodia. Thus, the cell extends forward towards the growth factor and grasps the ECM. Proteases released in response to the growth factor stimulus cut through the ECM to allow the cell to penetrate the matrix. Release of focal adhesions in the rear (where there is less or no growth factor) results in amoeboid movement of the cell up the growth factor gradient.

11 Intracellular Signals (Signal Transduction) that Regulate Cellular Events

Growth factor receptors are decision switches that translate extracellular signals to initiate cellular activities. Each receptor has a defined specificity for certain growth factors. The receptor is activated when it binds its ligand, the growth factor. The ability of a receptor to bind a growth factor is expressed in terms of its affinity, which in turn is expressed mathematically, in terms of the free growth factor concentration, as a dissociation constant (K_d). The higher the affinity, the lower the K_d and the tighter the binding between receptor and growth factor.[3] The lower the K_d, the more sensitive the cell will be to the presence of a particular growth factor. Growth factor receptors generally have K_ds in the picomolar (10^{-12}) or high femtomolar (10^{-15}) range. The in vivo concentrations of growth factors are very difficult to determine. But, because the K_d of the receptor determines the concentration range of growth factor over which the activation level of the receptor changes in vivo, this number can be used to estimate the likely concentrations of growth factor that are present in vivo when the receptor is activated.

Most growth factor receptors are transmembrane proteins with three domains. The extracellular domain is the growth factor binding domain. The transmembrane domain is generally a short polypeptide chain that forms an

---

[3]In this context, sometimes the association constant, $K_a = 1/K_d$ may be used as a direct measure of binding affinity.

alpha helical structure. The intracellular domain is a type of enzyme called a protein kinase that, when activated by a growth factor, transfers the phosphate from ATP to either tyrosine (EGF, TGFα, PDGF, FGF, VEGF receptors) or serine and threonine (TGFβ receptors) side-chains on other proteins and on itself. Cytokine receptors and integrins also use protein kinases in their responses to ligand activation, but the protein kinase is not part of the receptor. One or more cytoplasmic protein kinases are activated when the receptor's extracellular domain binds to its cognate cytokine or when integrins bind to their ECM targets. In many cases the protein kinase(s) become associated with receptors as a result of changes in structure of the intracellular domains of the receptors by which new sites for protein interactions are created.

An active receptor consists of more than one receptor protein, each protein component of which is called a subunit. The active receptor can be a multimer of the same type of receptor subunits (EGF, FGF, VEGF receptors) called a homodimer or can be a multimer of different types of receptor subunits (TGFβ, IFNγ receptors) called a heterodimer. Those receptors that form homodimers often form heterodimers with other receptor monomers of the same family that are related in sequence and structure and that are expressed in the same cells. For example, there are three VEGFRs, VEGFR1 (also called FLT-1), VEGFR2 (also called KDR in humans and flk-1 in mice) and VEGFR3. Of these, at least VEGFR1 and VEGFR2 can form homodimers and heterodimers.

In some cases (integrins) both inactive and active receptors are dimers and ligand binding results in a change in structure within the dimer [121]. In other cases (EGF, TGFβ, FGF, VEGF, IFNγ receptors) the individual receptor subunits are believed to be distributed independently on the membrane and ligand binding increases their affinity for each other with the resulting formation of an active dimer. Once formed, the dimerized receptor can have a higher affinity for the ligand than the monomer, which stabilizes the dimeric structure. For example, the VEGFR2 dimer binds VEGF 100 times more tightly than does the monomer [30].

Many growth factors, including VEGF and FGF, are also dimers (Fig. 9). In their respective growth factor-receptor complexes the growth factor dimers interact with each receptor subunit of the receptor dimer. Heterodimeric growth factors can also sometimes form that promote the formation of certain heterodimeric receptors. For example, heterodimers of PLGF and VEGF subunits will cause the formation of VEGFR1 and VEGFR2 heterodimers because PLGF only binds to VEGFR1 and VEGF binds to VEGFR2. Receptor homodimers and heterodimers are likely to have different structures and thus may have different functions as seems true for the VEGFRs [68].

In the case of a dimeric growth factor that forms a tetrameric active receptor:growth factor complex the dependence of receptor activation on growth factor concentration is biphasic if the growth factor can bind to both receptor monomers. Initially, with increasing concentration of the dimeric growth factor, the receptor activation increases until all receptor subunits are involved in

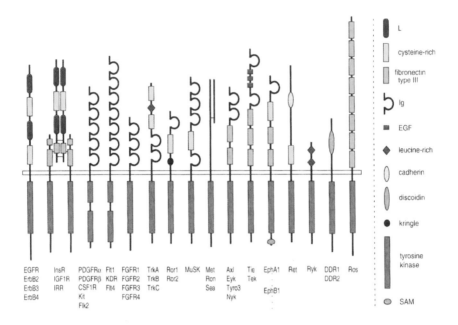

Fig. 8. Tyrosine kinase Receptors. The surface of a cell is a complex place. In this figure, classes of tyrosine kinase receptors are displayed, the cytosolic side being below the horizontal bar. Below each receptor is a set of names, each referring to a different receptor in the class indicated. For example, Flt1, KDR and Flt4 are all growth factor receptors of the VEGFR type. (From Hubbard and Till Annu Rev Biochem 2000; 69:373–98). Reprinted, with permission, from the Annual Review of Biochemistry, Volume 69 © 2000 by Annual Reviews www.anualreviews.org

tetrameric complexes consisting of one receptor dimer and one growth factor dimer. As the growth factor concentration increases beyond this saturation point trimeric complexes of growth factor dimers with receptor monomers become increasingly common with the resulting decrease in the number of active receptor:growth factor tetramers. This phenomenon has been observed for FGF and VEGF receptor activation profiles and cellular responses that involve receptor-growth factor complexes in which growth factor can bind each receptor monomer but not for cellular responses to TGFβ where only one of the heterodimeric receptor subunits has a significant affinity for the TGFβ dimer [30, 124].

The plasticity of protein structure results in a change in the overall structure of the receptor dimer within the receptor:growth factor complex when the growth factor and receptor subunits interact. Thus, growth factors, cytokines and ECM ligands activate their cognate receptors by changing the structural interfaces of interaction between individual receptor subunits and thereby changing the structure of the receptor. This structural change in the receptor:growth factor complex is transmitted across the body of the receptor

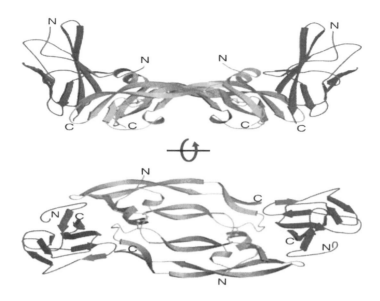

Fig. 9. VEGF-receptor binding. "A ribbon diagram ... with the two protomers of disulfide-linked VEGF shown in orange and purple, and Ig-like domain 2 of Flt1 shown in green. The view in the bottom panel is orthogonal to that in the top panel, as indicated." From Hubbard and Till Annu Rev Biochem 2000; 69:373–98 Fig. 3. Reprinted, with permission, from the Annual Review of Biochemistry, Volume 69 © 2000 by Annual Reviews www.anualreviews.org

from outside to inside the cell where the intracellular domains, now structurally altered, are functionally activated.

In most cases, activation of receptor function is associated with phosphorylation of the receptor or of associated proteins by the receptor kinase domain or by associated protein kinases. Phosphorylation alters protein structure and function. Thus, proteins phosphorylated by growth factor receptors and by cytokine receptor- and integrin-associated protein kinases can often interact differently with other proteins to alter their intracellular location, protein associations, and/or activity, which in turn changes the impact of these proteins on cellular function.

When the intracellular domain of a receptor is modified by phosphorylation it also becomes a binding site for many cytoplasmic proteins that contain specific domains (called SH2 domains) that recognize the phosphorylated receptor amino acid side chain and its surrounding structure. These interactions bring other proteins that interact with the SH2-domain proteins close to the receptor and promote their phosphorylation. When a growth factor receptor activates a protein by phosphorylation a domino effect is often initiated, which involves a cascade of events collectively called a signal transduction pathway. (Figs. 7 and 10). Each signal transduction pathway involves a different set of

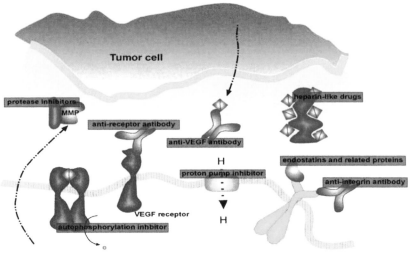

Fig. 10. Potential sites for inhibiting tumor induced angiogenesis.Many inhibitors of angiogenesis are being tested in clinical trials. These include protease inhibitors, antibodies against VEGF, the VEGF receptor, or the integrins, inhibitors of the autophosphorylation (activation) of the VEGF receptor, heparin-like drugs that soak up the VEGF, endostatins and related angiogenesis inhibitors, and inhibitors of certain general cell functions associated with angiogenesis such as inhibitors of proton pumps

proteins and can include proteins that bind to other proteins, enzymes such as protein kinases that phosphorylate other proteins, transcription factors that regulate gene expression, and proteins that regulate each of the previously listed activities.

Most signal transduction pathways have more than one molecular target and thus alter more than one cellular function. Cellular functions that can be altered by activated receptors include 1) enzymes such as the metabolic enzymes that provide energy to the cell, 2) structural proteins such as the protein components of the cytoskeleton that form the cell's shape and the proteins that form the focal adhesion complexes that determine where the cell will attach to the ECM, 3) transcription factors that regulate the expression of particular genes such as those required for passage through the cell cycle, and 4) proteins or enzymes that alter the distribution and quantity of other proteins within the cell such as proteins that are released by the cells, exposed in the cell surface or move from the cytoplasm to the nucleus to alter DNA synthesis or gene activity.

Most receptors activate more than one signal transduction pathway. Some of these signal transduction pathways target the same molecular species or target two molecules that interact either physically or by virtue of their molecular targets and their effects on cell function. The interaction of one signal transduction pathway with another to alter the functional outcome is called cross-talk. Depending on the nature of the molecular targets and the effect of the activated signal transduction pathway on them, the result of activating two signal transduction pathways simultaneously can be more than additive (e.g. synergistic) or can cancel individual pathway effects on a particular cell function [15, 81, 111, 112].

Each cell type expresses a different complement of genes that defines them. Thus, ECs and pericytes are differentiated from each other and from other cell types by the set of genes that are active (expressed) in these cells. Genes that encode receptors are amongst the genes that define a cell type. The receptors exposed to the surface will determine the extracellular signals to which the particular cell can respond and will therefore determine the signal transduction pathways that can be activated in that cell. In turn, expression of the genes that encode the protein components of the signal transduction pathways will determine which signal transduction pathways are activated in a particular cell. Also, the presence or absence of different molecular targets (primarily determined by their gene expression) will determine which cellular functions are altered by the signal transduction pathways and how these functions are altered.

Yet another impact on cellular response can be effected by the expression of genes that alter the intracellular location of a particular protein or that alter the half-life of the protein. The protein products of certain genes can also alter the intracellular locations of certain protein components of signal transduction cascades. If the component of the signal transduction cascade is not located appropriately in the cell, the cascade will not be activated even though the gene is expressed and the protein is present in the cell. Similarly, if a protein is synthesized but rapidly degraded, its concentration will be low and potentially limiting for the signal transduction pathway or the cellular response in which it participates. Thus, the response of ECs to their environment that results in angiogenesis is a combinatorial function of a large number of molecular signal interactions inside and outside the cells that involve many proteins including growth factors, receptors, signal transduction molecules and target molecules inside and outside the cells.

12 How does one Model These Events Mathematically?

12.1 Discrete Verses Continuum Models-a Question of Scale

Currently there is no mathematical model that attempts to include all the chemical and cellular components in one large master set of differential equations. Moreover the necessary complexity of such a model would undoubtedly limit its usefulness. Furthermore, it is clear from the forgoing discussion that the processes involved in angiogenesis are not completely understood at the biochemical/biological level. Therefore the mathematical modeling of this process is somewhat like shooting at a moving target. As the biology develops, so must the modeling. Conversely, and certainly far more interestingly, can the model predict testable hypotheses?

Over the years a number of authors, including the authors of this article have developed various simplified models. We refer the reader to www.ncbi.nlm.nih.gov/entrez/querey.fcgi where keyword searches will yield several dozen articles on the subject.

In this section we present an overall approach to modeling angiogenesis based strictly on biochemical kinetics and continuum mechanics. That is, the model we discuss here is a population model, one that looks at the movements of large numbers of cells. Such models are often called continuum models, in contrast to models which follow the movement of individual cells. In a rough sense, population models are rather like quantum mechanics, where one takes the point of view that electrons are probability densities, rather than as in classical mechanics, where one views them as individual particles.

12.2 The Mathematical Ideas in Prospective

The model we propose here is a dynamical system. That is, it is a system of ordinary and partial differential equations (pdes) in the space-time domain. The pdes appear on first inspection, to be parabolic, and indeed, each single equation is. However, those involving cell movement via chemotaxis or haptotaxis are strongly coupled. Thus, they not only possess a hyperbolic character, but also the character of mixed type equation.

In order to understand this in the simplest case, consider the system $u_t = du_{xx} - (uv_x)x$, $v_t = \epsilon v_x x + u - av$, the model of Keller-Segal (where $a, d\epsilon$ are non negative.) Suppose all three constants are positive. The first equation is parabolic u while, when $\epsilon > 0$ the second is parabolic in v. If $d = 0$ the first equation becomes hyperbolic in u. If $\epsilon = 0$ and we take $d > 0$, and we eliminate $u = v_t + av$ from the first equation, we have

$$v_{tt} + v_{tx}v_x + [v_t + a(v - d)]v_{xx} + av_t = v_{txx} - av_x^2 \ .$$

Ignoring the third order term for the moment, the second order operator on the left hand side has discriminant $v_x^2 - 4[v_t + a(v - d)]$ which can change sign.

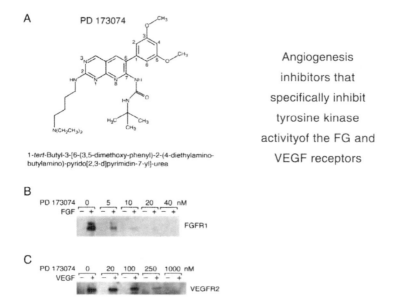

A   PD 173074

1-*tert*-Butyl-3-[6-(3,5-dimethoxy-phenyl)-2-(4-diethylamino-butylamino)-pyrido[2,3-d]pyrimidin-7-yl]-urea

Angiogenesis
inhibitors that
specifically inhibit
tyrosine kinase
activityof the FG and
VEGF receptors

B   PD 173074   0   5   10   20   40 nM
         FGF   − + − + − + − + − +

FGFR1

C   PD 173074   0   20   100   250 1000 nM
        VEGF   − + − + − + − + − +

VEGFR2

Mohammadi et al. 1998 EMBO J 17,5896-5904

Fig. 11. Growth factor receptor blocking by inhibition of tyrosine kinase activity

This means that the equation for v is of mixed type. (The third order term v_{txx} can be viewed as a strong damping term.) See [16, 41, 55, 59, 76, 77, 83] for various mathematical results concerning this system.

12.3 A Quick Review of Enzyme Kinetics and the Michealis-Menten Hypothesis

Suppose that we have a chemical reaction to convert S to and P. This may be represented symbolically as

$$S \leftrightarrow P$$

and which is energetically favorable, i.e. there is a net loss of free energy for the conversion of the substrate S to the product P. Such a reaction is said to be thermodynamically favorable or spontaneous. In many cases, there is an energy barrier between the two states that prevents the reaction from proceeding. However, a catalyst can sometimes be added to this system that lowers this barrier to such a degree as to make the reaction kinetically possible by speeding up the arrival to equilibrium by several orders of magnitude. For example, the conversion of CO_2 (carbon dioxide) to H_2CO_3 (carbonic acid) in water is accelerated by a factor of 10^6 in each direction by the enzyme *carbonic anydrase*. (A catalyst cannot change the thermodynamics, i.e. the ultimate ratio of the concentrations of products to reactants. It can only change the speed of the reaction in each direction.[4]) When the catalyst is a protein, it is called an enzyme and its name ends in "ase". The kinetic mechanism proposed by Micheialis and Menten by which enzymes catalyze such reactions takes place in two steps. First the enzyme binds to the substrate:

$$E + S \underset{k_{off}}{\overset{k_{on}}{\rightleftarrows}} \{ES\} .$$

Then there is conversion of the intermediate to product and release of the enzyme:

$$\{ES\} \overset{k_m}{\to} E + P .$$

The intermediate molecular species $I = \{ES\}$, will be more likely to convert to products P than to revert to substrate S. (The product need not be a single molecular unit. This is especially true when the conversion of S to P involves the cleavage of one or more of the peptide bonds in S.

Such mechanisms lie at the heart of many signal transduction pathways. They may be further regulated by competitive or non competitive inhibition. (In the former case, a second substrate S' competes with S for E via $S' + E \to J \to$ **no reaction** whereas in the latter, the intermediate is inhibited, i.e. $S' + \{ES\} \to J \to$ **no reaction**.)[5]

A word about notation: Chemists generally denote the concentration (in moles or micro moles per unit volume) of species A with square brackets vis: [A]. In most cases, systems are considered to be well stirred so that [A] depends at most on time. However, we need to assume that the concentrations of chemical species also depend on position. Therefore we sometimes abandon the brackets notation and write $a(x, y, z, t)$ or $a(\cdot, t)$ when we consider the concentration as a point function.

In so far as mass action alone is concerned, the above system yields, in the well stirred situation, a system of four kinetic ordinary differential equations:

$$
\begin{aligned}
\frac{d[S]}{dt} &= -k_{on}[E][S] + k_{off}[\{ES\}] , \\
\frac{d[E]}{dt} &= -k_{on}[E][S] + (k_{off+k_m})[\{ES\}] , \\
\frac{d[\{ES\}]}{dt} &= k_{on}[E][S] - (k_{off+k_m})[\{ES\}] , \\
\frac{d[P]}{dt} &= k_m[\{ES\}] .
\end{aligned}
\tag{1}
$$

---

[4]However, by either increasing the concentrations of reactants or decreasing the concentrations of products via other reactions, this ratio may be changed.

· [5]In many biological systems, such inhibitions are sometimes reversible. For example, $\{ES'\} + S \leftrightarrows \{ES\} + S'$ so that excess substrate S can overcome the inhibitory effects of S'. From the point of view of the inhibition of tumorigenic angiogenesis irreversible inhibition is perhaps more desirable.

In the well stirred situation with substrates and enzyme with relatively long half lives, biochemists and applied mathematicians have devoted considerable energy to understanding this system from their respective points of view. See [75] for an excellent discussion of this from the mathematician's viewpoint in this well stirred case.

If $[E]_0$ denotes the initial concentration of enzyme, the sum of the second and third of these equations tells us that in the well stirred case, $[E](t) + [\{ES\}](t) = [E]_0$. *However, in vivo, this is usually not the case, as the enzyme may be the product of another reaction pair, be degraded by virtue of having a short half life or be inactivated by binding to other proteins or by being sequestered from its substrate; for example located in a different cell compartment such as the nucleus or mitochondrion.*

Biochemists generally assume that such mechanisms are of "Michealis-Menten" type. This means that the enzyme substrate complex ($\{ES\}$) is assumed to come to equilibrium on a time scale that is much shorter than that required for the complete conversion of the substrate (S) into the product (P). From a mathematical point of view this says that the left hand side of the third equation in (1) is vanishes, i.e.

$$[\{ES\}] = \frac{k_{on}}{k_{off} + k_m}[E][S] = [E][S]/K_M \tag{2}$$

where $K_m = \frac{k_{off}+k_m}{k_{on}}$ is called the Michealis constant. (A very nice discussion of this hypothesis in terms of the language of singular perturbation theory (the matching of inner and outer solutions) is given in [75].) We shall assume this condition in the following without further ado. We see that the condition tells us the enzyme concentration is also constant. This is unrealistic, but to be expected because once the hypothesis is invoked we are dealing with the outer solution of the system (1). See [75]. Applying (2) to the conservation law, we are led to a single ordinary differential equation for the consumption of substrate:

$$\frac{d[S]}{dt} = -\frac{k_m[E]_0[S]}{K_m + [S]} \ .$$

Once $[S]$ is found, $[P]$ may be found by quadrature.

12.4 The Role of Chemical Kinetics in the Simplification of the Intracellular Events

In the simplest model of angiogenesis, the endothelial cells that line a given capillary are induced to proliferate and migrate by growth factor that has been secreted by a tumor or a gland. In order for such cells to migrate into the surrounding ECM and begin to build a new capillary, three events must occur. First, the capillary lining and the surrounding tissue must be degraded. Second, the cells must be capable of sensing and responding to this degraded state. Finally, new cells must form to fill the void left by moving cells.

A simplified model for this can be described as follows:

1. A molecule of growth factor, G, binds to a cell receptor, R to form an intermediate, $\{GR\}$ which in turn initiates an intracellular signal cascade which results in:
    a. The expression by the cell of one or more molecules of a proteolytic enzyme C and
    b. initiation of the cell cycle.
2. The protease breaks down the collagen matrix F, which results in a number of smaller peptides, $P.P'\ldots$, at least one of which may have an inhibiting effect on the process by inhibiting G, R, C or by blocking the further production of C by binding with $\{GR\}$. (This is a kind of "negative feedback loop.")
3. Inhibitors may be introduced into the system (Fig. 10).

In order to keep track of the bookkeeping, we need to recognize several states for the proteins that will do the work, R, G, C. If a molecule X is in an active state, we denote it by X_a otherwise it is in an inhibited state, X_i.

In terms of chemical equations, [1a] can be summarized as

$$G_a + R_a \overset{k_1}{\underset{k_{-1}}{\rightleftarrows}} \{G_a R_a\} ,$$

$$Y + \{G_a R_a\} \overset{k_2}{\rightarrow} nC_a + G' + R_a \tag{3}$$

where Y denotes the cell resources used during the cell cycle that are assembled to manufacture the protease, C_a in the active state. Here G' denotes the degradation products of growth factor. Some of the molecules of C are in a configuration so as to catalyze the breakdown of the extracellular matrix:

$$C_a + F \overset{k_3}{\underset{k_{-3}}{\rightleftarrows}} \{C_a F\} ,$$

$$\{C_a F\} \overset{k_4}{\rightarrow} P + P' + C_a . \tag{4}$$

The Michaelis-Menten hypothesis is to be in force for both (3)–(4), i.e.

$$[\{G_a R_a\}] = [G_a][R_a]/K_m^1 ,$$
$$[\{C_a F\}] = [C_a][F]/K_m^2 . \tag{5}$$

If we have a competitive inhibitor present, it can interfere with the receptor, the growth factor or the enzyme (Fig. 10). That is, at least one of

$$R_a + I_r \overset{\nu_r}{\rightleftarrows} R_i ,$$
$$G_a + I_g \overset{\nu_g}{\rightleftarrows} G_i ,$$
$$C_a + I_c \overset{\nu_c}{\rightleftarrows} C_i \tag{6}$$

must be in force. *These inhibitors may come from an external source, be secreted by the cells, or be found from among the products of matrix degradation, i.e., among $P, P', \ldots$.*

We can write down the "conservation laws" for the receptor, growth factor and protease densities as:

$$[R] = [R_a] + [R_i] + [\{G_a R_a\}] \,,$$
$$[G] = [G_a] + [G_i] + [\{G_a R_a\}] \,,$$
$$[C] = [C_a] + [C_i] + [\{C_a F\}] \,.$$

The equilibria (6) give rise to

$$\nu_r [R_a][I_r] = [R_i] \,,$$
$$\nu_g [G_a][I_g] = [G_i] \,,$$
$$\nu_c [C_a][I_c] = [C_i] \,.$$

However, it the inhibitor is noncompetitive, then one of

$$\{G_a R_a\}_a + \hat{I}_r \overset{\hat{\nu}_r}{\rightleftarrows} \{G_a R_a\}_i \,,$$
$$\{G_a R_a\}_a + \hat{I}_g \overset{\hat{\nu}_g}{\rightleftarrows} \{G_a R_a\}_i \,, \qquad (7)$$
$$\{C_a F\}_a + \hat{I}_c \overset{\hat{\nu}_c}{\rightleftarrows} \{C_a F\}_i$$

must be in force where now the " ^ " denotes non-competitive equilibrium. Then the conservation laws:

$$[\{G_a R_a\}] = [\{G_a R_a\}_a] + [\{G_a R_a\}_i] \,,$$
$$[\{C_a F\}] = [\{C_a F\}_a] + [\{C_a F\}_i]$$

will be in force along with the equilibrium equations:

$$\hat{\nu}_r [\{G_a R_a\}_a][\hat{I}_r] = [\{G_a R_a\}_i] \,,$$
$$\hat{\nu}_g [\{G_a R_a\}_a][\hat{I}_g] = [\{G_a R_a\}_i] \,,$$
$$\hat{\nu}_c [\{C_a F\}_a][\hat{I}_c] = [\{C_a F\}_i] \,.$$

The questions of which molecules are inhibited and what type of inhibition is in play (Fig. 10).

The answers depend, to a degree, on the molecular geometry that comes into play as well as on the mechanisms involved in the reactions.

When we write down the mass action rate laws, we must also take into account the sources of these inhibitors. One obvious source is that they are introduced intravenously. This was the point of view taken in [58, 60]. However, inhibitors, as we remarked above, arise as fragments of collagen decay among other processes.

For simplicity, let us consider the case in which growth factor activates a cell receptor while the enzyme that results from it degrades the collagen matrix. Furthermore, suppose that among the products of this degradation is a competitive inhibitor of growth factor, I_g. Assume also that we are no longer in the "well stirred" situation. Then chemical equations become:

$$G_a + R_a \underset{k_{-1}}{\overset{k_1}{\rightleftharpoons}} \{G_a R_a\} \overset{k_2}{\rightarrow} C_a + R_a \,,$$

$$C_a + F \underset{k_{-3}}{\overset{k_3}{\rightleftharpoons}} \{c_a F\} \overset{k_4}{\rightarrow} C_a + I_g + F' \,, \qquad (8)$$

$$G_a + I_g \overset{\nu_g}{\rightleftharpoons} G_i \,.$$

If we write down all eight differential equations coming from the first two of these chemical equations via mass action and use the Michealis-Menten hypothesis **without recourse to conservation laws** we obtain:

$$\frac{\partial g_a}{\partial t} = (-k_1 + k_{-1}/K_m^1)g_a r_a = -\frac{k_2 g_a r_a}{K_m^1} \,,$$

$$\frac{\partial c_a}{\partial t} = \frac{k_2 g_a r_a}{K_m^1} \,,$$

$$\frac{\partial f}{\partial t} = (-k_3 + k_{-3}/K_m^2)c_a f = -\frac{k_4 c_a f}{K_m^2} \,, \qquad (9)$$

$$\frac{\partial i_g}{\partial t} = \frac{k_4 c_a f}{K_m^2} \,.$$

We must also allow for molecular decay of G_a, C_a, I_g and tie these rates to the rates of diffusion of G, C, I_g. Moreover, a term must be included that reflects the capacity for cell expression of collagen.

We have some choices for our set of dependent variables. If we chose $\{g, c, r, f, i_g\}$ as our set, we can find r_a, c_a, g_a from the equations that arise from the "conservation laws".

$$r_a = \frac{r}{1 + g_a/K_m^1} \,,$$

$$g_a = \frac{g}{1 + \nu_g i_g + r_a/K_m^1} \,, \qquad (10)$$

$$c_a = \frac{c}{1 + f/K_m^2} \,.$$

With this choice we can include in the diffusion equations for g, c the kinetic rate terms coming from the reaction mechanisms (9).

It is also necessary to have a movement equation for receptor density. If ρ is the number of receptors of the type under consideration per cell, $r(x, yz, t)$ is the number of receptors in micro moles per liter and $N(x, y, z, t)$ is the cell density in cells per liter, then $\rho = r/N$ will be the number of receptors per cell expressed in micro moles per cell. If $1/N_{max}$ denotes the volume of a single cell, then $r_{max} = \rho N_{max}$ and thus $r = r_{max} N/N_{max}$. The quantity r_{max} may be estimated as follows: For many growth factors, there are roughly 10^5 receptors per endothelial cell. The volume of a typical EC is about 10^3 cubic microns or 10^{-15} cubic meters [78]. This means that there are roughly 10^{20} receptors per cubic meter or 10^{17} per liter [6, 7, 118]. Dividing this by Avogadro's number, 6×10^{23} we conclude that there are 1.2×10^{-6} moles per cubic liter or roughly 1.2 micromoles per liter. Thus we typically take $r_{max} \approx 1.0\mu M$.

Cells also express collagen. One way to model the expression of collagen is to use a logistic term for collagen production that also depends upon the relative cell density. that is, we include a term of the form $f(1 - f/f_M)(N/N_{max})$ where f_M is the density of pure collagen ($1/f_M$ is the specific volume of collagen.)

Before discussing the cell movement equations, we summarize the equations we have thus far in view of the forgoing comments. We let D_z, μ_z be the diffusion coefficient and the decay rate of molecular species z. (Decay rate $= \ln 2/(\text{half life})$.) We also let σ_z denote the rate at which cells can express z in micro molarity per unit time. We track the variables g, c via diffusion. If we do this, we must subtract from $\partial_t g$ the rate terms that correspond to the cellular consumption of active growth factor and the decay of active growth factor while adding a term that expresses the cellular expression of growth factor. A similar adjustment must be done for the protease rate equation. Then the rate laws become

$$
\begin{aligned}
\frac{\partial g}{\partial t} &= D_g \Delta g + \left(\sigma_g - \frac{k_2 g_a}{K_m^1 + g_a} \right) \frac{N}{N_{max}} - \mu_g g_a \\
\frac{\partial c}{\partial t} &= D_c \Delta c + \frac{k_2 g_a}{K_m^1 + g_a} \frac{N}{N_{max}} - \mu_c c_a , \\
\frac{\partial f}{\partial t} &= \frac{4}{T_f} f \left(1 - \frac{f}{f_M} \right) \frac{N}{N_{max}} - \frac{k_4 c_a f}{K_m^2} , \\
\frac{\partial i_g}{\partial t} &= D_{i_g} \Delta i_g + \frac{k_4 c_a f}{K_m^2} + \sigma_i \frac{N}{N_{max}} - \mu_{i_g} i_g
\end{aligned}
\tag{11}
$$

where we have written

$$
\frac{k_2 g_a r_a}{K_m^1} = \frac{k_2 g_a r}{K_m^1 + g_a} = \frac{k_2 g_a}{K_m^1 + g_a} \frac{N}{N_{max}}
$$

in view of the comment that $r_{max} \approx 1.0\mu M$. (Here $4/T_f$ is just a convenient way of writing the time scale for collagen production.) The equations (10) are then used to compute g_a, c_a. The form on the decay terms ($\nu_g g_a$ and $\mu_c c_a$) is taken to reflect that it is only the active form of the molecular species that undergoes "decay".

12.5 The Role of the Cell Cycle

For a given cell, we can expect that a receptor is either in an active state, ready to interact with growth factor, or in an inactive state. Furthermore, there are two subclasses of cells, those that are in the resting state G_0 and those that are in one of the other states (G_1, S, G_2, M). We denote by N_i the population of cells that are in the resting state. Then a simple logistic rate law[6] for the cells in one of the cell cycle states is given by (in the well stirred case):

$$\frac{dN}{dt} = \lambda \frac{g_a}{K + g_a} N \left(1 - \frac{N + N_i}{N_{max}}\right) - \mu N . \qquad (12)$$

It is perhaps important to emphasize the meaning of such a logistic equation from the point of view of the cell cycle. The equation itself is only a population equation. It does not tell us anything about which phase in the cell cycle a given cell is in. It is only a model of how we believe *the local population density* of cells moving through the states G_1, S, G_2, M is growing. However, it is not completely unrelated to the cell cycle in the sense that at saturation with a low cell density

$$\frac{dN}{dt} = (\lambda - \mu)N .$$

This tells us that the time through one pass of the cell cycle is $\ln 2/(\lambda - \mu)$.

Likewise, although individual cells in the G_0 state do not proliferate, in any given small region of space, the local density can change with time as cells pass in and out of this state. For example, as the capillary advances, the local population of cells at a fixed point behind the tip does grow from zero to some maximum number by virtue of entry of proliferating cells into the G_0 state. *We would like to have a description of the local population in this state that reflects the fact that as factors, which encourage the proliferation of active cells, decrease the population of cells in the G_0 state increases and vice versa.* That is, we would like a way to describe the rate of exit from the cell cycle as a function of the concentrations of the factors that encourage cell proliferation. We could model this by introducing a second inhibitor I_r that competes with growth factor for active receptors R_a.

Another approach is to argue that there is some logistic influence on the population of cells in the G_0 state that decreases as the concentration of growth factor increases vis:

$$\frac{dN_i}{dt} = \lambda \frac{K}{K + g_a} N_i \left(1 - \frac{N + N_i}{N_{max}}\right) \qquad (13)$$

where we have assumed that the apoptosis rate for quiescent cells is zero.[7]

---

[6]Equations of this form where the cells have two different carrying capacities have been used in [32, 33, 61].

[7]Here the formal "doubling time" is $\ln 2/\lambda$ for small populations of quiescent cells. Because the equation is a description of the growth of the quiescent cell population

Adding the last two equations and writing $N_T = N + N_i$ we have

$$\frac{dN_T}{dt} \leq \lambda \frac{(g_a N + K N_i)}{K + g_a} \left(1 - \frac{N_T}{N_{max}} \right) \tag{14}$$

so that (by the maximum principle) the total cell density cannot exceed the carrying capacity.

The idea here is that as the concentration of growth factor increases, the coefficient $\frac{g_a}{K+g_a}$ increases (i.e. the proliferation rate of active cells increases), while the coefficient $\frac{K}{K+g_a}$ decreases. It says that, as the growth factor concentration increases without bound, the rate of growth of the population of quiescent cells goes to zero. A model that would shut off the growth of the population of quiescent cells at a finite value of growth factor concentration can be obtained by replacing $\frac{K}{K+g_a}$ by $\frac{max\{K-\nu g_a, 0\}}{K+g_a}$ so that as soon as the concentration of growth factor exceeds K/ν the population density of quiescent cells stops increasing.

In some situations, the coefficient $\frac{g_a}{K+g_a}$ may be replaced by a function, $\varphi(g_a)$ which first increases and then decreases as g_a increases. Such a choice means that active cells would lose the ability to respond to growth factor if the concentration is large enough. The corresponding coefficient in the rest state equation would replaced by $\varphi_{max} - \varphi(g_a)$. Under either circumstance, the proliferation rates of cells in the rest state and cells in the active state move in opposition to one another.

In both equations (12), (13) there are no terms that allow for cell movement/chemotaxis/haptotaxis. We turn next to a modification of these equations that allows for this and that permits us to distinguish between the active cells and the resting cells even further. *It is important to emphasize here that the addition of diffusive and chemotactic terms to the logistic growth equations can markedly affect the dynamical behavior of solutions of such systems. Even the simple additional replacement of constants in the standard logistic equations by space or time dependent known functions will have similar effects. See for example, [11, 20, 44–46]*

12.6 The Role of Chemotaxis, Haptotaxis and Chemokinesis in Modeling Cell Movement

If **J** is the flux of a substance moving through a fluid, then mass balance leads us to the so called "equation of continuity"

$$\frac{\partial \eta}{\partial t} = -\nabla \cdot \mathbf{J}$$

density, this constant cannot be interpreted as a time through the cell cycle when the population density is small. The population density of quiescent cells increases because some cells enter into the G_0 state, not because they are dividing.

when there are no sources or sinks of the substance present. Here η is the concentration in mass or moles per unit volume of the substance while the flux carries the same mass or mole units but per unit area rather than per unit volume. In the case of ordinary Fickian diffusion (Brownian motion), the flux is related to the concentration by a constitutive equation known as Fick's Law:

$$J = -D\nabla\eta$$

where D is a coefficient of proportionality carrying the units of area per unit time. When the substance is molecular, it is called a diffusion coefficient. When we are dealing with large numbers of cells, we call it a cell movement coefficient. In the case that it is constant, eliminating the flux between the two equations above leads to the ordinary diffusion equation

$$\frac{\partial\eta}{\partial t} = D\Delta\eta$$

where $\Delta = \nabla\cdot\nabla = \nabla^2$ denotes the Laplace operator. This is the usual equation for small molecular species diffusing in an isotropic fluid such as water or benzene in the absence of convective flow and sources or sinks of this species.

Suppose the medium is not isotropic. Then D must be replaced by a (3×3) tensor that reflects the structure of the medium. In this case,

$$\mathbf{J} = -\mathfrak{D}(x,y,z)\nabla\eta$$

where $\mathfrak{D}$ may be position dependent. This is the case in an extracellular matrix that consists of cells, collagen and other connective tissue.

Further complicating the movement of our species, is the fact that its movement can be influenced by other species in the local environment. In the case of cell movement, this can happen because certain cell surface molecules can attach themselves to specific anchor points in the matrix that may either prevent motion (cell adhesion), or in the case of pseudopodia, encourage motion (haptotaxis). Or it could happen because the cell, through surface receptors, detects biomolecules that either encourage movement in the direction of the concentration gradient of that molecule or that encourage movement in the direction opposite to the concentration of that molecule. Either type of movement is called chemotaxis. (A further distinction is sometimes in that when the magnitude of the gradient response depends on the concentration of the molecule as well as on the magnitude of the gradient, the response is called chemokinetic.) If there are several agents present, such as growth factors, proteases, structure proteins, then the flux of cell density will be influenced by several gradients. For example, suppose that $g(\mathbf{x},t), c(\mathbf{x},t), f(\mathbf{x},t)$ are the concentrations of growth factor, protease and collagen and $\eta(\mathbf{x},t)$ is

$$\mathbf{J} = -D(\eta)\nabla_{\mathbf{x}}\eta + [G(\eta,g,c,f)\nabla_{\mathbf{x}}g + M(\eta,g,c,f)\nabla_{\mathbf{x}}c + F(\eta,g,c,f)\nabla_{\mathbf{x}}f]\eta$$

where G, M, F are some phenomenological functions of (η, g, c, f). These functions determine the influence of the specific species on the flux of the cell density. For example, at those points where $G > 0$, the flux of growth factor opposes the diffusion flux while where $G < 0$, the flux of growth factor assists the diffusion flux.

A further simplification of this flux vector can be made if we can assume that there is a potential function $T(g, c, f)$ such that $G = D(\eta)\partial_g T, M = D(\eta)\partial_c T, F = D(\eta)\partial_f T$, an assumption that holds if and only if $\text{curl}(G, M, F) = \overrightarrow{0}$.[8] Setting $\tau(g, c, f) = \exp(T(g, c, f)$, and using the resulting flux vector in the continuity equation we obtain:

$$\frac{\partial \eta}{\partial t} = \nabla \cdot \left\{ D(\eta) \left[\nabla \ln \left(\frac{\eta}{\tau(g, c, f)} \right) \right] \right\}. \tag{15}$$

Notice that it does not matter whether or not D is constant, a scaler or even a 3×3 matrix. The idea of [85] was to use the random walk argument of [17] together with the assumption of a constant diffusion coefficient to pass from a discrete random walk equation that results from the consideration of particle movement that is biased by chemotactic movement to a continuous equation in the particle density that reflects this bias. Here we obtain the continuous form of the chemotactic movement equation, (15) via elementary considerations.

The advantage of using (15) as a starting point for a cell movement equation becomes apparent if one assumes further that the biochemical species act independently of each other, i.e.,

$$\tau(g, c, f) = \tau_1(g)\tau_2(c)\tau_3(f).$$

We can determine the qualitative form of the individual factors based on our knowledge of how each agent influences cell motion. For example, it is known that as growth factor concentration increases, the chemotactic response first increases and then decreases. This is also true of protease and matrix protein. However, the relative sizes of these functions may be different and not all factors need be present in the response function. *Never-the-less, (15) dictates the rule of thumb: "Cell density follows the chemotactic response function."* The function τ is sometimes called the probability transition rate function (PTF) or the chemotactic response function.

12.7 Contact Inhibition and Crowding and How to Model Them

The presentation here was inspired by [42, 87]. Suppose that we have two cell types and we denote by η_1, η_2 their probability density functions (local concentrations). Suppose they move not only randomly but also "self chemotactically" and "chemotactically" in the sense that the flux of each species is determined not only by the gradient of the species itself but also by how

---

[8]when there is only one agent present, such a function T always exists.

the two species interact. For example the flux $\mathbf{J}_1$ of the first species takes the form:

$$\mathbf{J}_1 = -\mathcal{D}_{11}(\eta_1, \eta_2)\nabla_{\mathbf{x}}\eta_1 - \mathcal{D}_{12}(\eta_1, \eta_2)\nabla_{\mathbf{x}}\eta_2$$

where the cross term $\mathcal{D}_{12}(\eta_1, \eta_2)$ need not be of one sign. Thus, where this coefficient is positive, the (random) movement of the first cell type is aided by the gradient of the second type and where this coefficient is negative, the gradient of the second type tends to inhibit the movement of the first cell type.

A double application of the continuity equation for each species separately leads to the system

$$\partial_t\eta_1 = \nabla_{\mathbf{x}}[\mathcal{D}_{11}(\eta_1, \eta_2)\nabla_{\mathbf{x}}\eta_1] + \nabla_{\mathbf{x}}[\mathcal{D}_{12}(\eta_1, \eta_2)\nabla_{\mathbf{x}}\eta_2] \ .$$
$$\partial_t\eta_2 = \nabla_{\mathbf{x}}[\mathcal{D}_{21}(\eta_1, \eta_2)\nabla_{\mathbf{x}}\eta_1] + \nabla_{\mathbf{x}}[\mathcal{D}_{22}(\eta_2, \eta_2)\nabla_{\mathbf{x}}\eta_2] \ .$$
$$(16)$$

Suppose that there is a scalar function $\mathcal{D}_1(\eta_1, \eta_2)$ such that the vector field

$$[A_1, B_1] \equiv \left[\frac{\mathcal{D}_{11}(\eta_1, \eta_2)}{\eta_1 \mathcal{D}_1(\eta_1, \eta_2)}, \frac{\mathcal{D}_{12}(\eta_1, \eta_2)}{\eta_1 \mathcal{D}_1(\eta_1, \eta_2)}\right]$$

is exact in the triangular region defined by the inequalities $0 \le \eta_1, \eta_2$ and $0 \le \eta_1/N_1 + \eta_2/N_2 \le 1$ where N_i is the carrying capacity of species i. That is, there is a scaler function $\tau_1(\eta_1, \eta_2)$ such that $\nabla[\ln(\tau_1)] = [A_1, B_1]$, i.e.

$$\mathcal{D}_{11}(\eta_1, \eta_2) = \mathcal{D}_1(\eta_1, \eta_2)\{1 - \eta_1\partial_{\eta_1} \ln[\tau_1(\eta_1, \eta_2)]\} \ ,$$
$$\mathcal{D}_{12}(\eta_1, \eta_2) = -\mathcal{D}_1(\eta_1, \eta_2)\eta_1\partial_{\eta_2} \ln[\tau_1(\eta_1, \eta_2)] \ .$$

Suppose also that there are corresponding functions $\mathcal{D}_1, \tau_2$ such that

$$\mathcal{D}_{22}(\eta_1, \eta_2) = \mathcal{D}_2(\eta_1, \eta_2)\{1 - \eta_2\partial_{\eta_2} \ln[\tau_2(\eta_1, \eta_2)]\} \ ,$$
$$\mathcal{D}_{21}(\eta_1, \eta_2) = -\mathcal{D}_2(\eta_1, \eta_2)\eta_2\partial_{\eta_1} \ln[\tau_2(\eta_1, \eta_2)] \ .$$

Then (16) can be written in the more compact forme:

$$\partial_t\eta_1 = \nabla_{\mathbf{x}}\left\{\mathcal{D}_1(\eta_1, \eta_2)\eta_1\nabla_{\mathbf{x}} \ln\left[\frac{\eta_1}{\tau_1(\eta_1, \eta_2)}\right]\right\} \ ,$$
$$\partial_t\eta_2 = \nabla_{\mathbf{x}}\left\{\mathcal{D}_2(\eta_1, \eta_2)\eta_2\nabla_{\mathbf{x}} \ln\left[\frac{\eta_2}{\tau_2(\eta_1, \eta_2)}\right]\right\} \ .$$
$$(17)$$

Notice once again the ubiquitous logarithm.

As a simple example, suppose for $i = 1, 2$ $\tau_i(\eta_1, \eta_2) = \eta_i[1 - (\eta_1/N_1 + \eta_e/N_2)]$ and that the $\mathcal{D}_i$ are constant. (Here $1/N_i$ is the specific volume of species i.) Then the dynamics suggest that η_i will follow $\eta_i[1 - (\eta_1/N_1 + \eta_e/N_2)]$. Consequently, with no flux boundary conditions, the quantity $(\eta_1/N_1 + \eta_e/N_2)$ should evolve to a constant. Notice that, when η_{3-i} is fixed, τ_i first increases and then decreases as η_i increases from 0 to N_i. In other words, there is a tendency for the cells to first to aggregate and then to disaggregate near the carrying capacity.

It is perhaps worth mentioning that such a reduction is not, in general, always possible because the conditions for exactness may not be satisfied for any $\mathcal{D}_i$ in the region of interest in the $\eta_1 - \eta_2$ plane. That is, the first order partial differential equation for $\mathcal{D}_i$, $\partial_{\eta_2} A_i = \partial_{\eta_1} B_i$, may not be solvable in this region. The possibility of such a reduction is even less likely in three space dimensions (i.e. in real tissues) in the case that the coefficients $\mathfrak{D}_{ij}$ are tensors, i.e., the flux vectors are not coplanar with the cell density gradients.

Now we are in a position to modify (12), (13) to include cell movement. We assume

1. Active cells, i.e. cells in G_1, S, G_2, M are capable of expressing growth factor and collagen as described above.
2. Cells in G_0, the cells in the rest state do not express growth factor.
3. Active cell movement is induced by growth factor, protease and collagen gradients and contact inhibition.
4. Quiescent cell movement is induced solely by collagen gradients and contact inhibition.
5. The cell movement coefficients D_1, D_2 are constant.
6. There is a positive threshold for growth factor, g_e and transfer rates, δ, δ' between active and inactive cells such that when the growth factor is above this threshold, inactive cells become active at rate δ and when it is below this threshold, active cells become inactive at rate δ'.

This leads to the system:

$$\frac{\partial N}{\partial t} = \nabla \cdot \left\{ D_1 N \left[\nabla \ln \left(\frac{N}{\tau_1(N, N_i, g_a, c_a, f)} \right) \right] \right\}$$
$$+ \lambda \frac{g_a}{K + g_a} N \left(1 - \frac{N + N_i}{N_{max}} \right) - \mu N$$
$$+ \delta H(g - g_e) N_i - \delta' H(g_e - g) N ,$$

$$\frac{\partial N_i}{\partial t} = \nabla \cdot \left\{ D_2 N_i \left[\nabla \ln \left(\frac{N_i}{\tau_2(N, N_i, f)} \right) \right] \right\} + \lambda \frac{K}{K + g_a} N_i \left(1 - \frac{N + N_i}{N_{max}} \right)$$
$$- \delta H(g - g_e) N_i + \delta' H(g_e - g) N ,$$

(18)

$$\frac{\partial g}{\partial t} = D_g \Delta g + \left(\sigma_g - \frac{k_2 g_a}{K_m^1 + g_a} \right) \frac{N}{N_{max}} - \mu_g g_a ,$$

$$\frac{\partial c}{\partial t} = D_c \Delta c + \frac{k_2 g_a}{K_m^1 + g_a} \frac{N}{N_{max}} - \mu_c c_a ,$$

$$\frac{\partial f}{\partial t} = \frac{4}{T_f} f \left(1 - \frac{f}{f_M} \right) \frac{N + N_i}{N_{max}} - \frac{k_4 c_a f}{K_m^2} ,$$

$$\frac{\partial i_g}{\partial t} = D_{i_g} \Delta i_g + \frac{k_4 c_a f}{K_m^2} + \sigma_i \frac{N + N_i}{N_{max}} - \mu_{i_g} i_g .$$

The active protease and active growth factor are related to g, c by $g = g_a(1 + \nu_g i_g + r_a/K_m^1)$ and $c = c_a(1 + f/K_m^2)$ where $r_a = r_{max}(N/N_{max})/(1 + g_a/K_m^1)$. The first and third of these lead to $g = g_a(1 + \nu_g i_g + r_{max}(N/N_{max})/(g_a + K_m^1)) \equiv F(g_a)$. Typically, one assumes that $g_a \ll K_m^1$ so that the relation between g and g_a is linear. However, the function $F(\cdot)$ is strictly increasing, unbounded on $[0, \infty)$ and vanishes at $g_a = 0$. Therefore the resulting quadratic in g_a always has a unique positive solution and the approximation is unnecessary.

12.8 Other Housekeeping Chores-Model Extensions

No system such as (18) can be solved computationally as is. We need to close the system in a number of ways. We also need to describe the procedure for gleaning numerical information from the system. Currently the authors and M. W. Smiley are examining these issues. David Pinkston, an undergraduate honors student at Iowa State, has studied the one dimensional version of this system.

1. The geometry of the problem must be set. This is done in conjunction with a renormalization of all the variables to non dimensional variables. In the case of in situ angiogenesis or vasculogenesis, a bounded two or three dimensional region will suffice. However, if one wishes to describe the penetration of a nascent capillary into the surrounding ECM, it may be necessary to prescribe transmission boundary conditions from one domain to a second, adjacent domain. An illustration of this was carried out in [58] for a single cell type where Folkman's classic diagram for incipient angiogenesis [29] was used as a model. (The remaining boundary conditions can be set as dictated by the biological problem. For example, no flux boundary conditions or mixed type conditions as needed can be selected.)
2. Initial conditions for the six dynamic variables (N, N_i, g, c, f, i_g) must be set. These can be found by prescribing (N, N_i, g, c, f, i_g) initially. The primary difficulty here is in deciding how the initial densities of the quiescent and active cells are to be distributed. However, these may not always be known. We can proceed in one or more of three ways to test the model.
   a. One seeks spatially homogeneous steady states and studies, at least computationally, what happens when these steady states are perturbed.
   b. One can begin with homogeneous steady state condition as initial data and then drive the growth factor or inhibitor equations via source terms. For example, two such possibilities are:

$$\frac{\partial g}{\partial t} = D_g \Delta g + \left(\sigma_g - \frac{k_2 g_a}{K_m^1 + g_a}\right)\frac{N}{N_{max}} - \mu_g g_a + G_S(\mathbf{x}, t),$$
$$\frac{\partial i_g}{\partial t} = D_{i_g}\Delta i_g + \frac{k_4 c_a f}{K_m^2} + \sigma_i \frac{N}{N_{max}} - \mu_{i_g} i_g + I_S(\mathbf{x}, t). \tag{19}$$

One views G_S as an additional source of growth factor, perhaps from a small source within the region as from a tumor (a source function with compact support within the region of interest). The source term for inhibitor I_S can be viewed as coming from a remote tumor or as being introduced mechanically from the outside, much as one would introduce drugs intravenously.

3. These sources could be introduced via inhomogeneous terms in the boundary conditions for the fluxes of g, i on the boundary. This was the choice made in [58].

4. Other cell types can be included in this model along with the corresponding chemistry. A first attempt at including more than one cell type in angiogenesis modeling was carried out for the one dimensional problem in [60] where the roles of pericytes and macrophages were included. This was also the first time, to our knowledge that inhibitors were included in biochemical equations for angiogenesis modeling.

13 Vocabulary

Like most well developed disciplines, biochemistry and cell biology are rich in acronyms, jargon and terminology that seem impenetrable to the outsider. Likewise, mathematics possesses its own jargon, almost equally impenetrable to the non mathematician. Here we have included some definitions (some of which are taken from [1]).

aliphatic: referring to an organic molecule consisting only of carbon and hydrogen and having no closed chain structure.

amino acid: One of the twenty-one molecules that form the building blocks of proteins. They have an amino (basic) end, $-NH_2$ and a carbonyl (acidic) end $-COOH$

angiogenesis: The growth of new blood vessels from existing vessels.

apoptosis: Programmed cell death. It results in cells fragmenting into smaller lipid enclosed units.

ATP, adenosine triphosphate, a molecule consisting of one of the bases, adenine, a sugar and three phosphates. This molecule provides the energy needed to drive most cellular functions.

Cdk: cyclin dependent kinase.

Chemokine: A cytokine that stimulates directional cell migration of white blood cells.

CXC cytokines: A family of cytokines characterized by an amino acid sequence motif that includes CXC, where the letter C represents cysteine and X, refers to any aliphatic amino acid.

CXCL9: A chemokine induced by interferon.

CXCL10: Also called IFN-inducible protein 10 (IP10), a chemokine.

cytokine: Any of a class of immunoregulatory substances that are secreted by cells of the immune system.

ELR motif: motifs characterized by an amino acid sequence ELR (glutamic acid-leucine-arginine) that immediately precedes the first cysteine amino acid residue in the protein sequence.

ECM: The extracellular matrix of proteins that holds the cells of a tissue in place. Sometimes it refers to the tissue exterior to the circulatory system.

EGF, epidermal growth factor: A growth factor that stimulates the growth of epidermal cells.

endothelial cells (EC): The cells lining the interior of all blood vessels.

enzyme: A protein that functions as a catalyst to accelerate a reaction.

epitope: A molecular region on the surface of an antigen capable of eliciting an immune response and combining with the specific antibody produced by such a response.

FGF, fibroblast growth factor: A growth factor that stimulates the growth of fibroblasts (cells that maintain the ECM).

FLT-1, fms-like tyrosine kinase 1: The VEGF receptor-1.

growth factors: Any of a number of proteins that are capable of stimulating cell mitosis.

HSPG: heparan sulphate proteoglycan.

IFN, interferon: Any of a family of heat stable, soluble, basic, antiviral glycoproteins of low molecular weight usually produced by cells exposed to the action of virus or another intracellular parasite (as a bacterium) or experimentally to the action of chemicals.

integrins: A family of glycoproteins that are found on all or most mammalian cell surfaces that are composed of two dissimilar polypeptide chains.

-kinase: Any catalyst that aids in the addition of a phosphate group.

KDR, kinase insert domain-containing receptor: The human VEGF receptor-2, called flk-1 ("fetal liver kinase 1") in mice.

lumen: the central cavity of a tubular like structure, e.g. the channel through which the blood flows or the interior of the ureter. The bore of a needle.

MMP, matrix metalloprotease: One of a large family of zinc ion bearing proteases.

PAI: plasminogen activator, a protease inhibitor.

PDGF: platelet derived growth factor.

PEDF: pigment epithelium-derived factor

peptide bond: The bond that links the carbonyl carbon of one amino acid in a polypeptide protein to the adjacent nitrogen atom in the amine end of its nearest amino acid neighbor.

peptides: Short chains of amino acids.

phosphatase: Any catalyst that aids in the removal of a phosphate group.

PLGF: placental growth factor.

pro-uPA The inactive proform of the urokinase plasminogen activator that can be activated by cleavage.

proteoglycan: One of a number of high molecular weight glycoproteins found in the extracellular matrix of connective tissue, made up mostly of carbohydrate consisting of various polysaccharide side chains linked to a protein and resembling polysaccharides rather than proteins in their properties.

proteolysis: The cleavage of a peptide bond between two adjacent amino acids in a protein into two smaller peptides and a molecule of water.

RGD: The arginine-glycine-aspartate, amino acid sequence. A term used to define a domain (RGD domain) that is found in fibronectin and other ECM proteins that is bound by the integrin receptors.

proteolytic enzymes, proteases Enzymes that facilitate proteolysis.

sFLT-1: The soluble extracellular domain of the VEGF receptor-1.

TGFβ: Transforming growth factor type β, a growth factor for some cells and a growth inhibitor for others.

TIMP: Tissue inhibitor of metalloproteinases.

TNFα: Tumor necrosis factor α.

VEGF: Vascular endothelial cell growth factor.

uPAR: Urokinase plasminogen activator receptor.

uPA: Urokinase plasminogen activator, a protease.

References

1. *Medical Desk Dictionary*, Merriam Webster, 1996.
2. A. Abdollahi, P. Hahnfeldt, C. Maercker, H. Grone, J. Debus, W. Ansorge, J. Folkman, L. Hlatky, and P. Huber, *Endostatin's antiangiogenic signaling network*, Mol. Cell, 13 (2004), pp. 649–63.
3. B. Annabi, E. Naud, Y. Lee, N. Eliopoulos, and J. Galipeau, *Vascular progenitors derived from murine bone marrow stromal cells are regulated by fibroblast growth factor and are avidly recruited by vascularizing tumors*, J. Cell. Biochem., 2004 (2004), pp. 1146–58.
4. T. Asahara, C. Bauters, L. Zheng, S. Takeshita, S. Bunting, N. Ferrara, J. Symes, and J. Isner, *Synergistic effect of vascular endothelial growth factor and basic fibroblast growth factor on angiogenesis in vivo, Circulation*, 92 (1995), pp. II365–71.
5. P. Baluk, S. Morikawa, A. Haskell, M. Mancuso, and D. McDonald, *Abnormalities of basement membrane on blood vessels and endothelial sprouts in tumors*.
6. S. Batra and K. Rakusan, *Geometry of the vascular system-evidence for a metabolic hypothesis*, Microvasc. Res., (1991), pp. 39–50.
7. S. Batra and K. Rakusan, *Capillary network geometry during postnatal growth in rat hearts*, Am. J. Physiology, (1992), pp. 635–640.
8. R. Benezra and S. Rafii, *Endostatin's endpoints-deciphering the endostatin antiangiogenic pathway*, Cancer Cell, 5 (2004), pp. 205–6.
9. G. Bou-Gharios, M. Ponticos, Rajkumar, V., and D. Abraham, *Extra-cellular matrix in vascular networks*, Cell Prolif., 37 (2004), pp. 207–20.
10. F. Braet and E. Wisse, *Structural and functional aspects of liver sinusoidal endothelial cell fenestrae: a review*, Comp. Hepatol., 1 (2002), p. 1.

11. S. Cantrell, C. Cosner, and Y. Lou, *Multiple reversals of competitive dominance in ecological models via external habitat degradation*, J. Dynamic Diff. Eqns, in press (2005).

12. T. Chan-Ling, *Glial, vascular, and neuronal cytogenesis in whole-mounted cat retina*, Microsc. Res. Tech., 36 (1997), pp. 1–16.

13. M.A.J. Chaplain and A.M. Stuart, *A model mechanism for the chemotactic response of endothelial cells to tumor angiogenesis factor*, I.M.A. J. Math. Appl. Math. Biol., (1993), pp. 149–168.

14. E. Chen, S. Hermanson, and S. Ekker, *Syndecan-2 is essential for angiogenic sprouting during zebrafish development*, Blood, 103 (2004), pp. 1710–9.

15. C. Chiang and M. Nilsen-Hamilton, *Opposite and selective effects of epidermal growth factor and human platelet transforming growth factor-beta on the production of secreted proteins by murine 3t3 cells and human fibroblasts*, J. Biol. Chem., 261 (1986), pp. 10478–81.

16. S. Childress and J.K. Percus, *Nonlinear aspects of chemotaxis*, Math. Biosci., 56 (1981), pp. 217–237.

17. B. Davis, *Reinforced random walks*, Probal. Theory Related Fields, (1990), pp. 203–229.

18. P. DiMilla, J. Stone, J. Quinn, S. Albelda, and D. Lauffenburger, *Maximal migration of human smooth muscle cells on fibronectin and type iv collagen occurs at an intermediate attachment strength*, J. Cell Biol., 122 (1993), pp. 729–37.

19. V. Djonov, O. Baum, and P. Burri, *Vascular remodeling by intussusceptive angiogenesis*, Cell Tissue Res., 314 (Vascular remodeling by intussusceptive angiogenesis), pp. 107–17.

20. J. Dockery, V. Hutson, K. Mischaikow, and M. Pernarowski, *The evolution of slow dispersal rates: a reaction diffusion model*, J. Math. Biol., 37 (1998), pp. 61–83.

21. S. Egginton and M. Gerritsen, *Lumen formation: in vivo versus in vitro observations*, Microcirculation, 10 (2003), pp. 45–61.

22. S. Ekholm, P. Zickert, S. Reed, and A. Zetterberg, *Accumulation of cyclin e is not a prerequisite for passage through the restriction point*, Mol. Cell Biol., 21 (2001), pp. 3256–65.

23. R. Erber, A. Thurnher, A. Katsen, G. Groth, H. Kerger, H. Hammes, M. Menger, A. Ullrich, and P. Vajkoczy, *Combined inhibition of vegf and pdgf signaling enforces tumor vessel regression by interfering with pericyte-mediated endothelial cell survival mechanisms*, Faseb. J., 18 (2004), pp. 338–40.

24. A. Estreicher, J. Muhlhauser, J. Carpentier, L. Orci, and J. Vassalli, *The receptor for urokinase type plasminogen activator polarizes expression of the protease to the leading edge of migrating monocytes and promotes degradation of enzyme inhibitor complexes*, J. Cell. Biol., 111 (1990), pp. 783–92.

25. M. Fannon, K. Forsten-Williams, C. Dowd, D. Freedman, J. Folkman, and M. Nugent, *Binding inhibition of angiogenic factors by heparan sulfate proteoglycans in aqueous humor: potential mechanism for maintenance of an avascular environment*, Faseb. J., 17 (2003), pp. 902–4.

26. D. Fawcett, *A Textbook of Histology*, 1994.

27. R. Flaumenhaft, M. Abe, P. Mignatti, and D. Rifkin, *Basic fibroblast growth factor-induced activation of latent transforming growth factor beta in endothelial cells: regulation of plasminogen activator activity*, J. Cell Biol., 118 (1992), pp. 901–9.

28. J. Folkman, *Tumor angiogenesis: therapeutic implications.*, N. Engl. J. Med., 285 (1971), pp. 1182–6.
29. ——, *Angiogenesis-retrospect and outlook*, in Angiogenesis: Key Principles-Science-Technology-Medicine, P.B. Steiner, R. Weisz and R. Langer, eds., 1992.
30. G. Fuh, B. Li, C. Crowley, B. Cunningham, and J. Wells, *Requirements for binding and signaling of the kinase domain receptor for vascular endothelial growth factor*, J. Biol. Chem., 273 (1998), pp. 11197–204.
31. M. Furusato, M. Fukunaga, Y. Kikuchi, S. Chiba, K. Yokota, K. Joh, S. Aizawa, and E. Ishikawa, *Two- and three-dimensional ultrastructural observations of angiogenesis in juvenile hemangioma*, Virchows Arch. B Cell. Pathol. Incl. Mol. Pathol,, 46 (1984), pp. 229–37.
32. R.A. Gatenby and E.T. Gawlinski, *A reaction-diffusion model of cancer invasion*, Cancer Research, 56 (1996), pp. 5745–5753.
33. R.A. Gatenby and E.T. Gawlinski, *The glycolytic phenotype in carcinogenesis and tumor invasion: insights through mathematical models*, Cancer Research, 63 (2003), pp. 3847–3854.
34. E. George, H. Baldwin, and R. Hynes, *Fibronectins are essential for heart and blood vessel morphogenesis but are dispensable for initial specification of precursor cells*, Blood, 90 (1997), pp. 3073–81.
35. E. George, E. Georges-Labouesse, R. Patel-King, H. Rayburn, and R. Hynes, *Defects in mesoderm, neural tube and vascular development in mouse embryos lacking fibronectin*, Development, 119 (1993), pp. 1079–91.
36. R. Gerli, L. Ibba, and C. Fruschelli, *Ultrastructural cytochemistry of anchoring filaments of human lymphatic capillaries and their relation to elastic fibers*, Lymphology, 24 (1991), pp. 105–12.
37. A.J. Gospodarowicz, D. and J. Schilling, *Isolation and characterization of a vascular endothelial cell mitogen produced by pituitary-derived folliculo stellate cells*, Proc. Natl. Acad. Sci. USA, 86 (1989), pp. 7311–5.
38. E. Gustafsson and R. Fassler, *Insights into extracellular matrix functions from mutant mouse models*, Exp. Cell Res., 261 (2000), pp. 52–68.
39. M. Hangai, N. Kitaya, J. Xu, C. Chan, J. Kim, Z. Werb, S. Ryan, and P. Brooks, *Matrix metalloproteinase-9- dependent exposure of a cryptic migratory control site in collagen is required before retinal angiogenesis*, Am. J. Path., 161 (2002), pp. 1429–37.
40. K.M.L.P.A.A. Hellstrom, M. and C. Betsholtz, *Role of pdgf-b and pdgfr-beta in recruitment of vascular smooth muscle cells and pericytes during embryonic blood vessel formation in the mouse*, Development, 126 (1999), pp. 3047–55.
41. T. Hillen and P.A., *The one-dimensional chemotaxis model: global existence and asymptotic profile.*
42. T. Hillen and K. Painter, *Global existence for a parabolic chemotaxis model with prevention of overcrowding*, Adv. in Appl. Math., 26 (2001), pp. 280–301.
43. K. Hirschi and P. D'Amore, *Pericytes in the microvasculature*, Cardiovas. Res., 32 (1996), pp. 687–98.
44. V. Hutson, Y. Lou, and K. Mischaikow, *Spatial heterogeneity of resources versus Lotka-Volterra dynamics*, J. Differential Equations, 185 (2002), pp. 97–136.
45. V. Hutson, Y. Lou, K. Mischaikow, and P. Poláčik, *Competing species near a degenerate limit*, SIAM J. Math. Anal., 35 (2003), pp. 453–491 (electronic).
46. V. Hutson, K. Mischaikow, and P. Poláčik, *The evolution of dispersal rates in a heterogeneous time-periodic environment*, J. Math. Biol., 43 (2001), pp. 501–533.

47. R. Hynes, *Alteration of cell-surface proteins by viral transformation and by proteolysis*, Proc. Natl. Acad. Sci., 1973 (70), pp. 3170–4.

48. T. Ichii, H. Koyama, S. Tanaka, S. Kim, A. Shioi, Y. Okuno, E. Raines, H. Iwao, S. Otani, and Y. Nishizawa, *Fibrillar collagen specifically regulates human vascular smooth muscle cell genes involved in cellular responses and the pericellular matrix environment*, Circ. Res., 88 (2001), pp. 460–7.

49. D. Ingber and J. Folkman, *Inhibition of angiogenesis through modulation of collagen metabolism*, Lab. Invest., 59 (1988), pp. 44–51.

50. R. Iozzo and J. San Antonio, *Heparan sulfate proteoglycans: heavy hitters in the angiogenesis arena*, J. Clin. Invest., 108 (2001), pp. 349–55.

51. R. Jain, *Molecular regulation of vessel maturation*, Nat. Med., 9 (2003), pp. 685–93.

52. J. Jerdan, R. Michels, and B. Glaser, *Extracellular matrix of newly forming vessels–an immunohistochemical study*, Microvasc Res, 42 (1991), pp. 255–65.

53. F. Jonca, G. P. Ortega, N., N. Bertrand, and J. Plouet, *Cell release of bioactive fibroblast growth factor 2 by exon 6-encoded sequence of vascular endothelial growth factor*, J. Biol. Chem., 272 (1997), pp. 24203–9.

54. A. Jones, M. Hulett, J. Altin, P. Hogg, and C. Parish, *Plasminogen is tethered with high affinity to the cell surface by the plasma protein, histidine-rich glycoprotein*, J. Biol. Chem., 279 (2004), pp. 38267–76.

55. E. F. Keller and L. A. Segel, *Initiation of slime mold aggregation viewed as an instability*, J. Theor. Biol., 30 (1970), pp. 399–415.

56. G. R. B. W. A. K. T. R. F. R. Kostka, G. and M. Chu, *Perinatal lethality and endothelial cell abnormalities in several vessel compartments of fibulin-1-deficient mice*, Mol. Cell Biol., 21 (2001), pp. 7025–34.

57. T. Levchenko, K. Aase, B. Troyanovsky, A. Bratt, and L. Holmgren, *Loss of responsiveness to chemotactic factors by deletion of the c-terminal protein interaction site of angiomotin*, J Cell Sci, 116 (2003).

58. H.A. Levine, S. Pamuk, B.D. Sleeman, and M. Nilsen-Hamilton, *Mathematical modeling of capillary formation and development in tumor angiogenesis: Penetration into the stroma*, Bull. Math. Biol., (2001), pp. pp. 801–863.

59. H.A. Levine and B.D. Sleeman, *A system of reaction diffusion equations arising in the theory of reinforced random walks*, SIAM J. Appl. Math., (1997), pp. 683–730.

60. H.A. Levine, B.D. Sleeman, and M. Nilsen-Hamilton, *A mathematical model for the roles of pericytes and macrophages in the initiation of angiogenesis: I. the role of protease inhibitors in preventing angiogenesis*, Mathematical Biosciences, (2002), pp. 77–115.

61. H.A. Levine, M.W. Smiley, and M. Nilsen-Hamilton, *A mathematical model for the onset of avascular tumor growth in response to the loss of p53 function*, manuscript submitted.

62. S. Li, D. Harrison, S. Carbonetto, R. Fassler, N. Smyth, D. Edgar, and P. Yurchenco, *Matrix assembly, regulation, and survival functions of laminin and its receptors in embryonic stem cell differentiation*, 157 (2002), pp. 1279–90.

63. L. Liotta, P. Steeg, and W. Stetler-Stevenson, *Cancer metastasis and angiogenesis: an imbalance of positive and negative regulation.*, Cell, 64 (1991), pp. 327–36.

64. K. List, O. Jensen, T. Bugge, L. Lund, M. Ploug, K. Dano, and N. Behrendt, *Plasminogen-independent initiation of the pro-urokinase activation cascade in*

vivo. activation of pro-urokinase by glandular kallikrein (mgk-6) in plasmino-gendeficient mice., Biochemistry, 39 (2000), pp. 508–15.

65. X. Liu, H. Wu, M. Byrne, S. Krane, and R. Jaenisch, *Type iii collagen is crucial for collagen i fibrillogenesis and for normal cardiovascular development*, Proc. Natl. Acad. Sci. U.S.A., 94 (1997), pp. 1852–6.

66. J. Lohler, R. Timpl, and R. Jaenisch, *Embryonic lethal mutation in mouse collagen i gene causes rupture of blood vessels and is associated with erythropoietic and mesenchymal cell death*, Cell, 38 (1984), pp. 597–607.

67. M. Maragoudakis, E. Missirlis, G. Karakiulakis, M. Sarmonica, M. Bastakis, and N. Tsopanoglou, *sement membrane biosynthesis as a target for developing inhibitors of angiogenesis with anti-tumor properties*, Kidney Int., 1993 (1993), pp. 147–50.

68. T. Matsumoto and L. Claesson-Welsh, *Vegf receptor signal transduction*, Sci STKE, (2001), p. RE21.

69. A.M.L. Meyer, G.T., L. Noack, M. Vadas, and J. Gamble, *Lumen formation during angiogenesis in vitro involves phagocytic activity, formation and secretion of vacuoles, cell death, and capillary tube remodelling by different populations of endothelial cells*, Anat. Rec., 249 (1997), pp. 327–40.

70. J. Mezquita, B. Mezquita, M. Pau, and C. Mezquita, *Down-regulation of flt-1 gene expression by the proteasome inhibitor mg262*, J Cell Biochem, 89 (2003), pp. 1138–47.

71. J. Miner and P. Yurchenco, *Laminin functions in tissue morphogenesis*, Annu. Rev. Cell Dev. Biol., 20 (2004), pp. 255–94.

72. S. Mitola, M. Strasly, M. Prato, P. Ghia, and F. Bussolino, *Il-12 regulates an endothelial cell-lymphocyte network: effect on metalloproteinase-9 production*, J. Immunol., 171 (2003), pp. 3725-33.

73. T. Moser, D. Kenan, T. Ashley, J. Roy, M. Goodman, U. Misra, D. Cheek, and S. Pizzo, *Endothelial cell surface f1-f0 atp synthase is active in atp synthesis and is inhibited by angiostatin*, Proc Natl Acad Sci USA, 98 (2001), pp. 6656–61.

74. A. Murray, *Recycling the cell cycle: cyclins revisited*, Cell, 116 (2004), pp. 221–34.

75. J. Murray, *Mathematical Biology, Biomathematics Texts*, Springer-Verlag, 1989.

76. T. Nagai and T. Nakaki, *Stability of constant steady states and existence of unbounded solutions in time to a reaction-diffusion equation modelling chemotaxis*, Nonlinear Analysis, 58 (2004), pp. 657–81.

77. T. Nagai and T. Senba, *Global existence and blow-up of radial solutions to a parabolic-elliptic system of chemotaxis*, Adv. Math. Sci. Appl., 8 (1998), pp. 145–156.

78. R.M. Nerem, M.J. Levesque, and J.F. Cornhill, *Vasuclar endothelial cell morphology as an indicator of the pattern of blood flow*, J. Biomech. Eng., (1981), pp. 172–176.

79. A. Nerlich and E. Schleicher, *Identification of lymph and blood capillaries by immunohistochemical staining for various basement membrane components*, Histochemistry, 96 (1991), pp. 449–53.

80. G. Neufeld, T. Cohen, S. Gengrinovitch, and Z. Poltorak, *Vascular endothelial growth factor (vegf) and its receptors*, Faseb. J., 13 (1999), pp. 9–22.

81. M. Nilsen-Hamilton, R. Hamilton, W. Allen, and S. Potter-Perigo, *Synergistic stimulation of s6 ribosomal protein phosphorylation and dna synthesis by epidermal growth factor and insulin in quiescent 3t3 cells*, Cell, 31 (1982), pp. 237–42.

82. M. Nilsen-Hamilton, Y. Jang, M. Delgado, J. Shim, K. Bruns, C. Chiang, Y. Fang, C. Parfett, D. Denhardt, and R. Hamilton, *Regulation of the expression of mitogen-regulated protein (mrp; proliferin) and cathepsin l in cultured cells and in the murine placenta*, Mol. Cell Endocrinol., 77 (1991), pp. 115–22.

83. K. Osaki and A. Yagi, *Finite dimensional attractor for one-dimensional Keller-Segel equations*, Funkcial. Ekvac., 44 (2001), pp. 441–69.

84. L. Ossowski and J. Aguirre-Ghiso, *Urokinase receptor and integrin partnership: coordination of signaling for cell adhesion, migration and growth*, Curr. Opin. Cell. Biol., 12 (2000), pp. 613–20.

85. H. G. Othmer and A. Stevens, *Aggregation, blow up and collapse: The abc's of taxis and reinforced random walks*, SIAM J. Appl. Math., (1997), pp. 1044–1081.

86. U. Ozerdem and W. Stallcup, *Early contribution of pericytes to angiogenic sprouting and tube formation*, Angiogenesis, 6 (2003), pp. 241–9.

87. K. J. Painter and T. Hillen, *Volume-filling and quorum sensing in models for chemosensitive movement*, Can. J. Appl. Math., 10 (2003).

88. S. Palecek, J. Loftus, M. Ginsberg, D. Lauffenburger, and A. Horwitz, *Integrin-ligand binding properties govern cell migration speed through cell-substratum adhesiveness*, Nature, 385 (1997), pp. 537–40.

89. S. Pandey, C. Lopez, and A. Jammu, *Oxidative stress and activation of proteasome protease during serum deprivationinduced apoptosis in rat hepatoma cells; inhibition of cell death by melatonin*, Apoptosis, 8 (8), pp. 497–508.

90. B. Patterson and Q. Sang, *Angiostatin-converting enzyme activities of human matrilysin (mmp-7) and gelatinase b/type iv collagenase (mmp-9)*, J. Biol. Chem., 272 (1997), pp. 28823–5.

91. L. Pereira, K. Andrikopoulos, J. Tian, S. Lee, D. Keene, R. Ono, D. Reinhardt, L. Sakai, N. Biery, T. Bunton, H. Dietz, and F. Ramirez, *Targeting of the gene encoding fibrillin-1 recapitulates the vascular aspect of marfan syndrome*, Na Genet., 17 (1997), pp. 218–22.

92. X. Pi, Y. C., and B. Berk, *Big mitogen-activated protein kinase (bmk1)/erk5 protects endothelial cells from apoptosis*, Circ. Res., 94 (2004), pp. 362–9.

93. J. Pickering, C. Ford, and L. Tang, B. Chow, *Coordinated effects of fibroblast growth factor-2 on expression of fibrillar collagens, matrix metalloproteinases, and tissue inhibitors of matrix metalloproteinases by human vascular smooth muscle cells. evidence for repressed collagen production and activated degradative capacity*, Arterioscler. Thromb. Vasc. Biol., 17 (1997), pp. 475–82.

94. D. Qiao, K. Meyer, C. Mundhenke, S. Drew, and A. Friedl, *Heparan sulfate proteoglycans as regulators of fibroblast growth factor-2 signaling in brain endothelial cells. specific role for glypican-1 in glioma angiogenesis*, J. Biol. Chem., 278 (2003), pp. 16045–53.

95. M. Reyes, A. Dudek, B. Jahagirdar, L. Koodie, P. Marker, and C. Verfaillie, *Origin of endothelial progenitors in human postnatal bone marrow*, J. Clin. Invest., 109 (2002), pp. 337–46.

96. C. Ruegg and A. Mariotti, *Vascular integrins: pleiotropic adhesion and signaling molecules in vascular homeostasis and angiogenesis*, Cell Mol. Life Sci., 60 (2003), pp. 1135–57.

97. E. Ruoslahti and M. Pierschbacher, *New perspectives in cell adhesion: Rgd and integrins.*, Science, 238 (1987), pp. 491–7.

98. M. Rusnati and M. Presta, *Interaction of angiogenic basic fibroblast growth factor with endothelial cell heparan sulfate proteoglycans. biological implications in neovascularization*, Int. J. Clin. Lab. Res., 26 (1996), pp. 15–23.

99. F. Sabin, *Studies on the origin of the blood vessels and of red blood corpuscles as seen in the living blastoderm of chick during the second day of incubation*, Contr Embryol, 9 (1920), pp. 215–262.

100. T. Sasaki, H. Larsson, D. Tisi, L. Claesson-Welsh, E. Hohenester, and R. Timpl, *Endostatins derived from collagens xv and xviii differ in structural and binding properties, tissue distribution and anti-angiogenic activity*, J. Mol. Biol., 301 (2000), pp. 1179–90.

101. C. Scavelli, E. Weber, M. Agliano, T. Cirulli, B. Nico, A. Vacca, and D. Ribatti, *Lymphatics at the crossroads of angiogenesis and lymphangiogenesis*, J. Anat., 204 (2004), pp. 433–49.

102. J. Sechler and J. Schwarzbauer, *Control of cell cycle progression by fibronectin matrix architecture*, 273 (1998), pp. 25533–6.

103. G. Seghezzi, R.C. Patel, S., A. Gualandris, G. Pintucci, E. Robbins, R. Shapiro, A. Galloway, D. Rifkin, and P. Mignatti, *Fibroblast growth factor-2 (fgf-2) induces vascular endothelial growth factor (vegf) expression in the endothelial cells of forming capillaries: an autocrine mechanism contributing to angiogenesis*, J. Cell. Biol., 141 (1998), pp. 1659–73.

104. D. Seo, H. Li, L. Guedez, P. Wingfield, T. Diaz, R. Salloum, B. Wei, and W. Stetler-Stevenson, *Timp-2 mediated inhibition of angiogenesis: an mmp-independent mechanism*, Cell, 114 (2003), pp. 171–80.

105. D. Sims, *Diversity within pericytes*, Clin. Exp. Pharmacol. Physiol., 27 (2000), pp. 841–6.

106. J. Sottile, *Regulation of angiogenesis by extracellular matrix*, Biochim. Biophys. Acta, 1654 (2004), pp. 13–22.

107. J. Sottile, D. Hocking, and P. Swiatek, *Fibronectin matrix assembly enhances adhesion-dependent cell growth*, J. Cell. Sci., 111 (Pt 19) (1998), pp. 2933–43.

108. Y. Taba, M. Miyagi, M., H. Y., Inoue, F. Takahashi-Yanaga, S. Morimoto, and T. Sasaguri, *15-deoxy-delta 12,14-prostaglandin j2 and laminar fluid shear stress stabilize c-iap1 in vascular endothelial cells*, Am. J. Physiol. Heart Circ. Physiol., 285 (2003), pp. H38–46.

109. T. Tarui, M. Majumdar, L. Miles, W. Ruf, and Y. Takada, *Plasmin-induced migration of endothelial cells. a potential target for the anti-angiogenic action of angiostatin.*, PJ Biol Chem, 277 (2002), pp. 33564–70.

110. S. Taverna, G. Ghersi, A. Ginestra, S. Rigogliuso, S. Pecorella, G. Alaimo, F. Saladino, V. Dolo, P. Dell'Era, A. Pavan, G. Pizzolanti, P. Mignatti, M. Presta, and M. Vittorelli, *Shedding of membrane vesicles mediates fibroblast growth factor-2 release from cells.*, J. Biol. Chem., 278 (2003), pp. 51911–9.

111. F. Thalacker and M. Nilsen-Hamilton, *Specific induction of secreted proteins by transforming growth factor-beta and 12-o-tetradecanoylphorbol-13-acetate. Relationship with an inhibitor of plasminogen activator*, J. Biol. Chem., 262 (1987), pp. 2283–90.

112. ——, *Opposite and independent actions of cyclic amp and transforming growth factor beta in the regulation of type 1 plasminogen activator inhibitor expression*, Biochem. J., 287(3) (1992), pp. 855–62.

113. J. Thyboll, J. Kortesmaa, R. Cao, R. Soininen, L. Wang, S.L. Iivanainen, A., M. Risling, Y. Cao, and K. Tryggvason, *Deletion of the laminin alpha4 chain leads to impaired microvessel maturation*, Mol. Cell. Biol., 22 (2002), pp. 1194–202.

114. E. Tkachenko, E. Lutgens, R. Stan, and M. Simons, *Fibroblast growth factor 2 endocytosis in endothelial cells proceed via syndecan-4-dependent activation of rac1 and a cdc42-dependent macropinocytic pathway*, J. Cell Sci., 117 (2004), pp. 3189–99.

115. B. Troyanovsky, T. Levchenko, G. Mansson, O. Matvijenko, and L. Holmgren, *Angiomotin: an angiostatin binding protein that regulates endothelial cell migration and tube formation*, J. Cell Biol., 152 (2001), pp. 1247–54.

116. I. Vlodavsky, Z. Fuks, R. Ishai-Michaeli, P. Bashkin, E. Levi, G. Korner, R. Bar-Shavit, and M. Klagsbrun, *Extracellular matrix-resident basic fibroblast growth factor: implication for the control of angiogenesis*, J. Cell Biochem., 45 (1991), pp. 167–76.

117. E. Voest, B. Kenyon, M. O'Reilly, G. Truitt, R. D'Amato, and J. Folkman, *Inhibition of angiogenesis in vivo by interleukin 12*, J Natl Cancer Inst., 87 (1995), pp. 581–6.

118. J. Waltenberger, L. Claesson-Welsh, A. Siegbahn, M. Shibuya, and C.-H. Heldin, *Different signal transduction properties of kdr and flt1, two receptors for vascular endothelial cell growth factor*, J. Biolog. Chem., (1994), pp. 26988–26995.

119. H.R. Weber, I.T. and R. Iozzo, *Model structure of decorin and implications for collagen fibrillogenesis*, J. Biol. Chem., 271 (1996), pp. 31767–70.

120. S. Wickstrom, K. Alitalo, and J. Keski-Oja, *Endostatin associates with integrin alpha5beta1 and caveolin-1, and activates src via a tyrosyl phosphatase-dependent pathway in human endothelial cells*, Cancer Res, 62, pp. 5580–9.

121. T. Xiao, J. Takagi, B. Coller, J. Wang, and T. Springer, *Structural basis for allostery in integrins and binding to fibrinogen-mimetic therapeutics*, Nature, 432 (2004), pp. 59–67.

122. J. Xu, D. Rodriguez, E. Petitclerc, J. Kim, M. Hangai, Y. Moon, G. Davis, P. Brooks, and S. Yuen, *Proteolytic exposure of a cryptic site within collagen type iv is required for angiogenesis and tumor growth in vivo*, 2001.

123. S. Yamagishi, H. Yonekura, Y. Yamamoto, H. Fujimori, S. Sakurai, N. Tanaka, and H. Yamamoto, *Vascular endothelial growth factor acts as a pericyte mitogen under hypoxic conditions*, Lab. Invest., 79 (1999), pp. 501–9.

124. G. Zellin and A. Linde, *Effects of recombinant human fibroblast growth factor-2 on osteogenic cell populations during orthopic osteogenesis in vivo*, Bone, 26 (2000), pp. 161–8.

125. Z. Zhou, J. Wang, R. Cao, H. Morita, R. Soininen, K. Chan, B. Liu, Y. Cao, and K. Tryggvason, *Impaired angiogenesis, delayed wound healing and retarded tumor growth in perlecan heparan sulfate-deficient mice*, Cancer Res, 64 (2004), pp. 4699–702.

Mathematical Modelling of Proteolysis and Cancer Cell Invasion of Tissue

Georgios Lolas

mpokos@hotmail.com

Abstract. The growth of solid tumors proceeds through two distinct phases: the avascular and the vascular phase. It is during the latter stage that the insidious process of cancer invasion of peritumoral tissue can and does take place. Vascular tumors grow rapidly allowing the cancer cells to establish a new colony in distant organs, a process known as metastasis. The metastatic cascade is a multi-step process that involves the over-expression of proteolytic enzyme activity such as the urokinase-type plasminogen activator (uPA) and matrix metalloproteinases (MMPs). uPA initiates the activation of an enzymatic cascade that primarily involves the activation of plasminogen and subsequently its matrix degrading protein plasmin. Degradation of the peritumoral tissue enables the cancer cells to migrate through the tissue and subsequently to spread to secondary sites in the body.

In this chapter we consider a mathematical model of cancer cell invasion of extracellular matrix which focuses on the role of the urokinase plasminogen activation system. The model consists of a system of reaction-diffusion-taxis partial differential equations describing the interactions between cancer cells, urokinase plasminogen activator (uPA), uPA inhibitors, plasmin and the host tissue. That partial differential equations (naturally) focus on the spatio-temporal dynamics on the uPA system and its role in tissue invasion. The results obtained from the spatio-temporal systems, underscore the ability of rather simple models to produce complicated dynamics, associated with tumour heterogeneity and cancer cell progression and invasion.

1 Introduction

During the last two decades, mathematics has made a considerable impact as a useful tool with which to model and understand biological phenomena. Mathematical modelling is now recognised as an important part of understanding biomedical systems. Although there are obvious limitations which must be recognised, mathematics does indeed have a contribution to make, at least in helping us understand the very basic building blocks of behaviour exhibited by the cell, the gene, or the enzyme. Through mathematical modelling, biological phenomena of enormous complexity can be idealized and simplified in a set of mathematical equations.

G. Lolas: *Mathematical Modelling of Proteolysis and Cancer Cell Invasion of Tissue*,
Lect. Notes Math. **1872**, 77–129 (2006)
www.springerlink.com

This chapter considers a mathematical model in an attempt to understand certain properties or aspects of the urokinase plasminogen activation system and its role in cancer invasion and metastasis. However, we cannot model an entire biological system, and to ask the question of what role a specific molecular component plays in effecting and/or controlling the locomotive behaviour of a cell leads immediately to a need to deconvolute a highly complex system. Asking the next question of how to manipulate the effect of a specific component or to control its activity leads to the corresponding need to predict the outcome of a "reconvolution" of the system with altered properties. Last but not least, for the prediction of what will result from making any genetic or biochemical interventions targeting that component or other molecules that interact with it, it is necessary to integrate, or "reconstruct", the resulting alterations in the physical processes up to individual (stochastic) cell paths, and finally up to cell population (deterministic) distributions.

Before we continue we should make clear what form of equations will be analysed here, and why. The model considered here will focus on the level of cell populations or densities and their response to concentrations of chemicals. Under the continuum hypothesis, the spatio-temporal state of a system of cells and/or chemical interactions is described by partial differential equations (PDEs) derived from considerations of conservation of matter. Suppose we have a fixed but arbitrary volume V enclosed by a surface S and we consider the flow of cells through this volume. The conservation equation states that the rate at which the number of cells changes (accumulates or disappears) within V must be balanced by the net flow (influx or efflux) of cells across the bounding surface S, plus the number of mitotic, proliferating, and/or degrading cells in V, or

$$\frac{d}{dt} \int_V u(\mathbf{x}, t) \, d\mathbf{x} = \int_S -\mathbf{J}(\mathbf{x}, t) \cdot d\mathbf{S} + \int_V K(u, p) \, d\mathbf{V} , \tag{1}$$

where $u(\mathbf{x}, t)$ is the concentration of cells at position x and time t; $\mathbf{J}$ is the flux of cells through the closed surface, $\mathbf{S} = \partial V$, per unit volume per unit time; $K(u, p)$ describes the net rate of mitotic, proliferating, and/or degrading cells and is generally described by a polynomial or fractional function in u and p representing interactions with other cells and/or chemicals. Using the divergence theorem ($\int_V \nabla \cdot \mathbf{J} d\mathbf{V} = \int_S \mathbf{J} \cdot d\mathbf{S}$), (1) may be written

$$\frac{d}{dt} \int_V u(\mathbf{x}, t) \, d\mathbf{V} = \int_V (-\nabla \cdot \mathbf{J} + K(u, p)) \, d\mathbf{V} . \tag{2}$$

Assuming the domain is fixed in time, we may differentiate through the integral. Using the fact that the choice of volume V was arbitrary we have that at every point $(\mathbf{x}, t)$ the following conservation equation holds.

$$\frac{\partial u}{\partial t} = -\nabla \cdot \mathbf{J} + K(u, p) . \tag{3}$$

Systems of the above form have been used to model a wide variety of biological phenomena and a number of examples can be found in the books by Murray (2003) and Edelstein-Keshet (1988). A more formal derivation of the equation can be found in Okubo (1980).

Although over the past 30 years or so a number of mathematical models have been proposed in an attempt to describe various key stages of tumour development, up until now the development of a basic "consensus" model of solid tumour growth and development is still a mind-bending problem for existing and future mathematical biologists. In this regard, one of the major challenges of the next decade that mathematicians will face is to overcome Karlin's principle that overshadows every such model up until now, namely *"The purpose of models is not (necessarily) to fit the data, but to sharpen the questions"* (Karlin, 1983), and develop biologically realistic mathematical models which clarify fundamental cancer processes and which can predict new strategies of clinical therapy.

2 Mathematical Modelling of Solid Tumour Growth and Invasion

In vivo cancer growth is a complicated phenomenon involving many inter-related processes. Solid tumour growth occurs in two distinct phases, the initial growth being characterised as the avascular phase, the later growth as the vascular phase. The transition from avascular growth to vascular growth depends upon the crucial process of angiogenesis and is necessary for the tumour to attain nutrients and dispose of waste products (Folkman, 1974; 1976). To achieve vascularization, tumour cells secrete a diffusible substance known as tumour angiogenesis factor (TAF) into the surrounding tissues (Folkman and Klagsbrun, 1987). This has the effect of stimulating nearby capillary blood vessels to grow towards and penetrate the tumour, re-supplying the tumour with vital nutrient. Invasion and metastasis can now take place. By the time a tumour has grown to a size whereby it can be detected by, in the case of breast-cancer, simple self-examination, there is a strong likelihood that it has already reached the vascular growth phase. The primary aim of screening and the associated image enhancement technologies is therefore to detect cancers prior to this stage.

By contrast, tumour invasion is a relatively new area for mathematical modelling. However, over the last 10 years or so many mathematical models of tumour growth and invasion, both temporal and spatio-temporal, have appeared in the research literature. A noteworthy interplay between population ecological mathematical models and tumour biology was introduced by Gatenby (1995) and Gatenby and Gawlinski (1996). In his earlier paper Gatenby (1995) used a Lotka-Volterra competition model to examine tumour biology through the dynamic interaction of malignant and normal cells. Furthermore, in his latter paper, Gatenby and Gawlinksi (1996) considered a deterministic reaction-diffusion equation model for cancer invasion. A reaction-diffusion model was developed describing the spatio-temporal distributions of tumour and normal tissue as well as H^+ ion concentration. Finally, the mathematical model predicts that high H^+ ion concentrations in neoplastic tissue will extend by chemical diffusion, as a gradient into adjacent normal tissue, where normal cells are unable to survive in this acidic environment and this results in a progressive loss of layers of normal cells and thus tumour invasion evolves.

In a very comprehensive paper, Perumpanani et al. (1996) presented a theoretical model describing cell invasiveness as a function of tumour cell interactions with the local, normal host cells, noninvasive tumour cells, extracellular matrix proteins (ECM) and the proteases. Movement is described across a chemotactic/haptotactic gradient stimulus. Furthermore, their simulation studies demonstrated that the speed of invasiveness as well as the concomitant wave profile can be computed as a function of the tumour's phenotypic profile, its extracellular matrix make up, and the gradient stimuli the tumour finds itself in. In addition, their results highlighted the consequences of high protease production and excessive proteolysis of the extracellular matrix milieu in noninvasion.

The paper by Orme and Chaplain (1996) envisions a spherical tumour growing and invading with regard to the parent blood vessel vascularization which may consequently lead to metastasis. Their model describes the invasive tumour cells advancing towards the parent blood vessels (chemoattractants). However, capillary vessels were unable to reach some parts of the tumour (tumour centre) due to competition for space with tumour cells or high internal pressure which may cause vessels to collapse.

Turning the tables on traditional views on invasion, Perumpanani et al. (1998) suggested that extracellular matrix-mediated chemotaxis runs in the opposite direction to that of invasion. Briefly, the idea behind this concept is that during the process of human fibrosarcoma cell line (HT1080) migration, they showed that the degraded components of the extracellular matrix exert a chemotactic pull stronger than that of undigested fragments and that this runs in the opposite sense, against the direction of invasion.

In a more recent paper, Perumpanani and Byrne (1998) investigated whether regional variations in extracellular matrix concentration affect the propensity of tumours to invade a particular tissue. In other words, they predicted that for the fibrosarcoma cell line (HT1080) both directed movement (haptotaxis) up a collagen gradient as well as HT1080 cell proliferation are related to a collagen gel concentration in a biphasic manner. Of particular note is their assumption that protease production is proportional to the product of the tumour cell density and collagen gel concentration, as a consequence of signals transduced in the invading cells by the surrounding exctracellular matrix milieu or the collagen gel.

Modelling of a related phenomenon, embryonic implantation involving the invasion of trophoblast cells into maternal uterine tissue, using a deterministic reaction-diffusion approach, has also been carried out (Byrne et al., 1998). A novel feature of their model, is their assumption that trophoblast cells respond chemotactically to spatial gradients generated by the inhibitor, rather than the activator protease. Moreover, recently Byrne et al. (2001) presented a simpler submodel of the aforementioned model (Byrne et al., 1998), carrying out a mathematical analysis and obtaining a typical travelling wave solution of the submodel.

In a recent paper, Anderson et al. (2000) described a unifying conceptual theoretical framework for modelling tumour invasion and metastasis. They presented both deterministic and discrete approaches of describing the invasion of host tissue by tumour cells. The continuum approach examined the way that tumour cells respond to extracellular matrix gradients via haptotaxis in both one and two dimensions. In particular, the one dimensional model simulations highlight the possibility that a small cluster of cancer cells can easily secede from the primary body of the tumour as a result of random and biased migration as well as matrix degrading enzymes.

Furthermore, a pioneering contribution of the model lies in the fact that in their two dimensional results they consider the medium in which the tumour grows to be heterogeneous. By introducing extracellular matrix heterogeneity, cells are no longer clearly amassed in to those driven by haptotaxis and those driven by dispersion as a consequence of the already-existing gradients in the extracellular matrix. In addition, Anderson et al. (2000) developed an extended discrete model using as a basis the aforementioned continuum model. Even in the discrete model, they managed to confirm the importance of haptotaxis for both invasion and metastasis and they also underscore the effect of cell proliferation in invasion and migration of cancer cells as an eventuality of its space-filling function.

Additionally, Turner and Sherratt (2002), develop a discrete model of malignant invasion using an extension of the Potts model (Stott, 1999). In other words, they used the Potts model to simulate a population of malignant cells experiencing interactions due to both homotypic and heterotypic adhesion while also secreting proteolytic enzymes and experiencing a haptotactic gradient. In this regard, they investigated the influence of changes in cell-cell adhesion on the invasion process.

In summary, we note that deterministic reaction-diffusion equations have been used to model the spatial spread of tumours both at an early stage in its growth (Sherratt and Nowak, 1992) and at the later invasive stage (Orme and Chaplain, 1996; Gatenby, 1996; Perumpanani, 1996) while modelling of related phenomena, i.e. embryonic implantation involving invading trophoblasts cells, using a reaction-diffusion approach has also been carried out (Byrne et al., 1998).

However, we would like to emphasise that the models mentioned above consider the medium in which tumours grow to be homogeneous. On the contrary, in vivo tissues have a high degree of fine-scaled spatial structure and the paper by Anderson et al. (2000) investigates the effects of spatial heterogeneity. Mathematical models of cancer chemotherapy have also been developed to investigate, for example, the use of targeted enzyme-conjugates antibodies (ECA) for the selective activation of anti-cancer products (Jackson et al., 1999; 2000) and the effects of drug resistance on the optimal scheduling of drugs (Murray, 1997).

The aforementioned models have resulted in a novel perspective on the different stages of tumour growth. Mathematically, the models all essentially amount to systems of differential equations. However, it is known that the spatio-temporal dynamics of such biologically oriented models can be dramatically changed if we assume that the underlying environment to be spatially heterogeneous. In this regard, Maini et al. (1992) considered the problem of diffusion-driven instability in a system of reaction-diffusion equations. Their studies suggested that if we chose a bifurcation parameter (in this case, the ratio of chemical difussivities) to vary across a one-dimensional domain, the patterns exhibited by the system varied in amplitude and/or wavelength across the domain. Particularly, they suggested that if the parameter was chosen to be above its bifurcation value only in a sub-interval of the domain, then it was possible for patterns to propagate into the domain where linear analysis would predict stability of the uniform steady state – a situation they described as "environmental instability" (Benson et al., 1993). Based on these results we can deduce that environmental inhomogeneity could be an important regulator of biological pattern formation.

In line with the aforementioned suggestion of environmental heterogeneity, Swanson et al. (2000) considered tissue heterogeneity in the case of brain gliomas, which are generally highly diffuse. The impressive increased detection capabilities in computerised tomography (CT) and magnetic resonance imaging (MRI) have resulted in earlier detection of glioma tumours, although despite this progress the benefits of early treatment have been minimal. This is due to the fact that even after surgical excision well beyond the visible tumour boundary, regeneration near the edge of resection ultimately results. This is because the presently available imaging techniques only detect a small proportion of the actual, highly diffuse tumour.

Experiments in rats show that malignant gliomas cells implanted in rat brain disperse more quickly along white matter tracts than grey matter. Swanson (2000) considered a simple reaction-diffusion model for glioma cell invasion on a two-dimensional anatomically accurate slice of brain tissue in which they imposed a spatially dependent cell diffusion coefficient to account for different cell motility rates in grey and white matter. Using numerical simulation, they characterised how the proportion of a tumour that was detected depended on the cell diffusion coefficients and the cell proliferation rate. They also showed that the heterogeneity within the brain caused the dynamics of tumour invasion to vary significantly depending on the initial location of the tumour. These results have important implications on how much tissue a surgeon should aim to remove when a tumour is detected.

In this remaining sections of this chapter mathematical models describing the "kinesis", "taxis" and reactions of the urokinase plasminogen activation system (consisting of cancer cells, urokinase plasminogen activator (uPA) and extracellular matrix components i.e. vitronectin, fibronectin, laminin) are derived and developed to consider several key components of the system.

3 Biological Background

3.1 An Overview of Cancer

The word cancer is an "umbrella term" for approximately 200 diseases. Since the earliest medical records were kept, cancer as a disease has been described in the history of medicine. The origin of the word *cancer* is credited to the Greek physician Hippocrates (460–370 B.C.), considered the "Father of Medicine", who lifted medicine out of the realms of magic, superstition and religion.

Hippocrates used the terms καρκίνος (carcinos) and καρκίνωμα (carcinoma) (the ancient Greek word for "crab") to describe a group of diseases that he studied, including cancers of the breast, uterus, stomach and skin. The hard centre and spiny projections of the tumours as well as the tendency of tumours to reach out and spread that Hippocrates first observed reminded him very much of "the arms of a crab", because of the way a cancer adheres to any part of its surroundings that it seizes upon in an obstinate manner like the crab does.

Besides the popular generic term *"cancer"* that the English language has adopted (which is also the Latin word for crab), there is another technical medical term for cancer: νεοπλασία (neoplasia). Neoplasia or neoplasm literally means new (νέο) formation (πλάσις) in Greek. This indicates that cancers are actually new growths of cells in the body. In this regard the definition of a cancer is: *"a new growth of tissue resulting from a continuous proliferation of abnormal cells that have the ability to invade and destroy other tissues"* (King, 2000). Another term for cancer is "tumour". Tumour literally means "swelling" or mass. In this case, it refers to a mass of non-structured new cells, which have no known purpose in the physiological function of the body (Hanahan and Weinberg, 2000).

At the early growth stage the tumour is relatively harmless and is still *avascular*, that is, it lacks its own network of blood vessels for supplying nutrients, including oxygen, and for removing wastes (Folkman, 1974; 1976). The critical event that converts a self-contained pocket of aberrant cells into a rapidly growing malignancy comes when the tumour becomes *vascularized* (Folkman, 1976). That means that it has its own blood supply and microcirculation. A vascularized tumour has two distinct advantages over an avascular tumour:

(i) a direct supply of nutrients into the tumour. This results in a rapid increase in growth.

(ii) the tumour can shed cells into the bloodstream.

Normal, as well as neoplastic, tissues become vascularized by a process called angiogenesis, the growth of new capillary blood vessels from pre-existing vessels (Folkman, 1974; 1976; Pepper, 2001a). The first event of tumour-induced angiogenesis involves the secretion of a number of chemicals, collectively known as tumour angiogenic factors (TAF), into the surrounding tissue (Folkman and Klagsbrun, 1987). These factors diffuse through the tissue space creating a chemical gradient between the tumour and any existing vasculature. Upon reaching any neighbouring blood vessels, previously quiescent endothelial cells lining these vessels are stimulated to degrade their basement membrane, to invade the surrounding stroma and to migrate towards the tumour. As cells migrate, the endothelium begins to form sprouts which can then form loops and branches through which blood circulates. The whole process repeats forming a capillary network which eventually connects with the tumour, completing angiogenesis and supplying the tumour with the nutrients it needs to grow further. There is now also the possibility of tumour cells finding their way into the circulation and being deposited at distant sites throughout the body (Carmeliet and Jain, 2000).

3.2 Invasion and Metastasis

Cancers also possess the ability to actively invade the local tissue and then spread throughout the body. Invasion and metastasis are the most insidious and life-threatening aspects of cancer (Liotta and Stetler-Stevenson, 1991; Liotta and Clair, 2000). We first examine invasion. Whether physiological or malignant invasion, the regulation for its necessary events involves spatial and temporal coordination, as well as certain cyclic "on-off" processes, at the level of individual cells. Motility, coupled with regulated, intermittent adhesion to the extracellular matrix and degradation of matrix molecules, allows an invading cell to move through the three-dimensional tissue matrix. At the leading edge of the motile cell, receptor-ligand and proteolytic-antiproteolytic complexes coordinate sensing, protrusion, burrowing and traction of the cell (Liotta and Stetler-Stevenson, 1991; Lauffenburger and Horwitz, 1996).

Conventional wisdom used to hold that invasion (as well as metastasis) is a late event – often "too late" – in the clinical course of a patient's cancer. However, we now know that invasion can be both early and clinically "silent" (Aznavoorian et al., 1993). The threat of tumour invasion is exemplified by the fact that brain cancer does not need to metastasize to kill a patient. The growth of a brain tumour mass in the confined area of the skull causes compression damage: in addition, local invasion may enable brain tumour cells to move away from the primary tumour and to reach other sites within the brain. Such insidious behaviour may represent the inappropriate use of a programme responsible for the outgrowth of neuronal protrusions called neurites during normal neuronal development (Liotta and Clair, 2000). Indeed, cancer invasion in general may be a deregulated form of a physiological invasion process required for neuronal wiring in the embryo, tissue remodelling in the formation of blood vessels, and wound healing.

The most significant turning point in the disease (cancer), however, is the establishment of metastasis. The metastatic spread of tumour cells is the predominant cause of cancer deaths, and with few exceptions, all cancers can metastasize. At this stage, the patient can no longer be cured by local therapy alone. Metastasis is defined as the formation of secondary tumour foci at a site discontinuous from the primary tumour (Liotta and Stetler-Stevenson, 1991; Liotta and Clair, 2000). Metastasis unequivocally signifies that a tumour is malignant and this is in fact what makes cancer so lethal. In principal, metastases can form following invasion and penetration into adjacent tissues followed by dissemination of cells in the blood vascular system (hematogeneous metastasis) and lymphatics (lymphatic metastases). Sequential steps in the so-called *"metastatic cascade"* are believed to include the following:

- metastatic cells arise within a population of neoplastic/tumourigenic cells as a result of genomic instabilities;
- vascularization of the tumour through the angiogenesis process;
- detachment of metastatic-competent cells that have already evolved;
- migration of the metastatic cells;
- local invasion of cancer cells into the surrounding tissue, requiring adhesion to and subsequent degradation of extracellular matrix (ECM) components;
- transport of metastatic cells either travelling individually or as emboli composed of tumour cells (homotypic) or of tumour cells and host cells (heterotypic);
- metastatic cells survive their journey in the circulation system;
- adhesion/arrest of the metastatic cells at the secondary site, cells or emboli arrest either because of physical limitations (i.e. too large to traverse a lumen) or by binding to specific molecules in particular organs or tissues;
- escape from the blood circulation (extravasation);
- proliferation of the metastatic tumour cells;
- growth of the secondary tumour in the new organ.

Metastases can appear shortly after surgery but can also remain unde-tected for more than a decade before manifesting themselves clinically (King, 2000; Chambers et al., 2002; Fidler, 2002). This indicates that disseminated cancer cells can persist in a dormant state, unable to form a progressively in-creasing tumor mass (Chambers et al., 2002). Such heterogeneity of outcome indicates that the fate of tumour cells that disseminate to distant organs be-fore surgery must be regulated by either inherent cancer cell properties or the milieu of the target organs, or both. Identifying the mechanisms that keep metastases in their dormant, occult state is one of the most challenging and important avenues of cancer research (Chambers et al., 2002; Fidler, 2002).

It is well recognized that the majority of cells within a tumour cannot complete the process of metastasis. Indeed, a very small percentage (0.01%) of circulating tumour cells entering the bloodstream successfully form clinically detectable lesions (Fidler et al., 1991). Thus, metastasis is a highly selective competition, favouring the survival of a minor subpopulation of metastatic tumour cells that pre-exist within the primary tumour. By inference, it follows that a similarly small percentage of cells within a primary tumour would display a marker for metastasis (Duffy, 2001). A tumour marker can be defined as a substance produced either by a tumour or by the body in response to a tumour that aids cancer detection and/or monitoring. Just as it is easier to see a single, lit candle in a dark room than to find the only unlit candle in a room full of lit candles, it is easier to identify a single cell expressing a new metastatic marker against a background of non-expressing cells than it is to find non-expressing cells within a mass of cells that express a particular metastatic marker.

3.3 Tumour Heterogeneity

The development of metastasis is dependent on an interplay between host factors and intrinsic characteristics of malignant cells. The metastasis of cancer cells is one of the most devastating aspects of neoplasia. It is responsible for most therapeutic failures because patients succumb to multiple secondary tumour growths and not necessarily to the primary tumour. A major issue, which has recently received a great deal of attention, is the possibility that neoplasms are heterogeneous and contain subpopulations of cells with differing metastatic capabilities (Brattain et al., 1981). The question of heterogeneity in solid tumours has been of increasing interest to cancer biologists in recent years (Alexandrova, 2001). Therefore, we can ask: "Does the process of metastasis represent the random survival of tumour cells, or does it result from the survival and growth of a specialized subpopulation of cells ?".

During tumour evolution, genetic changes may lead to emergence of new tumour cell subpopulations with diverging phenotypic characteristics (Nowell, 1986; Fleuren, 1995), thus resulting in a heterogeneous mixture of cells differing in immunogenic properties, in metastatic ability and in responsiveness to cytotoxic agents etc. This process is a fundamental property of cancer and has important biological and clinical consequences among which augmentation of tumour progression and development of resistance to treatment are most crucial for the host. Not surprisingly then, it has also been demonstrated that neoplasms are also heterogeneous with regard to invasion and metastasis, i.e., that they contain a variety of subpopulations of cells with differing metastatic potentials (Fidler, 1978).

Tumours are variable in several ways. Their characteristics change with organ site and cell origin. Numerous host variables, such as age and hormonal status, also introduce differences. For example, the same cancer can even vary in different species as well as in commonly believed similar hosts. However, this inter-tumour variation is not what most investigators mean by "*tumour heterogeneity*". The term "**tumour heterogeneity**" means the existence of distinct subpopulations of tumour cells with specific characteristics within a single neoplasm (Heppner, 1984). However, even here, there is room for confusion. Tumours are architectually complex, differing regionally in vasculature, connective tissue components, and other characteristics which can alter the phenotype of otherwise identical cells. Marked differences in the proliferation behaviour of tumour cells within a single cancer cell are commonplace. Some cells, perhaps most, are reproductively dead; others are out of cycle; and still others are cycling but are, at a given time, at different stages in the process. Additionally, many cellular phenotypes, such as antigen expression, membrane composition, response to chemotherapy, metastatic proclivity, to name a few, are themselves functions of the cell cycle.

The possible existence of highly metastatic variant cells within a primary tumour suggests that we no longer should consider a neoplasm to be a uniform entity (Fidler, 1978; Heppner, 1984). In this regard, tumour heterogeneity and progression are major features of neoplastic development. Tumour cell societies are highly adapted for survival and proliferation. They can successfully survive natural and artificial (therapeutic) selection by producing new variants (Fidler, 1978; Heppner, 1984). Therefore, for cancer researchers and clinicians, tumour heterogeneity and progression represent problems that need to be solved or circumvented. A better understanding of tumour *anarchy* will help scientists to clarify such important biological phenomena such as appearance of metastases, drug resistance, spontaneous regression and could improve cancer prevention, diagnosis and therapy.

3.4 Proteolysis and Extracellular Matrix Degradation

The prognosis of a cancer is primarily dependent on its ability to invade and metastasize. Many steps that occur during tumour invasion and metastasis (as well as in a number of distinct physiological events in the healthy organism, such as trophoblast invasion, and skin wound healing) require the regulated turnover of extracellular matrix (ECM) macromolecules. A more localized degradation of matrix components is required when cells migrate through a basal lamina. It is now widely believed that the breakdown of these barriers is catalyzed by proteolytic enzymes released from the invading tumour. Most of these proteases belong to one of two general classes: many are metalloproteases while others are serine proteases (Andreasen et al., 1997; 2000). Proteases give cancers their defining characteristic – the ability of malignant cells to break out of tissue compartments.

However, proteolytic degradation of the extracellular matrix is essential for the processes of tissue remodelling as well. These processes take place in a number of distinct physiological events in the healthy organism, such as trophoblast invasion, mammary gland involution, and skin wound healing. The plasminogen activation system has an important position among the extracellular proteases engaged in these degradation reactions. This system is organized as a proteolytic cascade with active proteases and their pro-enzymes, protease inhibitors, and extracellular binding proteins.

Although it was originally thought that their role was simply to break down tissue barriers, enabling tumour cells to invade through stroma and blood vessel walls at primary and secondary sites, it is now understood that matrix metalloproteinases (MMPs) and plasminogen activators (PAs) also participate in angiogenesis (Pepper, 2001a) and are selectively upregulated in proliferating endothelial cells (Andreasen et al., 1997; 2000; Pepper, 2001a). In the following section, we will demonstrate the pleiotropic activities that the urokinase plasminogen activation system has in cell migration, cell movement, tumour progression, and metastasis. Rather than being comprehensive, the next section will cover specific areas which are currently undergoing rapid development.

3.5 The Urokinase Plasminogen Activation System: Biology and Regulation

Distant metastasis, and not the primary tumour itself, is the predominant cause of death in patients with malignant solid tumours. Thus, novel therapy concepts aimed at preventing tumour cell spread to distant organs are urgently needed. Up to now cancer drug development has focussed on the identification of molecules with cytotoxic activity against tumour cells. However, it has become evident that cytotoxic molecules, identified simply on the basis of their ability to poison as many cancer cells (and often, normal cells) as possible, are insufficient and, in some cases, undesirable to combat the progression of many tumours. In this regard, a paradigm shift is currently underway in the discovery of anti-cancer therapies focusing on the modulation of tumour characteristics other than tumour cell proliferation directly as a means of suppressing tumour growth, invasion and metastasis. These approaches include attempting to inhibit tumour neovascularization (angiogenesis), extracellular matrix (ECM) remodelling (e.g. during local invasion) and responsiveness of the tumour to growth factors as well as attempting to increase the rate of tumour cell apoptosis (Kohn and Liotta, 1995).

Extracellular proteolytic enzymes such as serine proteinases and metalloproteinases, have been implicated in cancer invasion, angiogenesis and metastasis, the basic idea being that the release of proteolytic enzymes in tumours facilitates cancer-cell invasion into the surrounding normal tissue by breakdown of basement membranes and ECM (see the review by Mignatti and Rifkin, 1993). It has been suggested that plasminogen activation is an essential prerequisite to many of the activities mediated by tumour-associated cells and it has been implicated in angiogenesis, growth factor activation and mobilization, ECM remodelling, invasion and metastasis goes back several decades (see the review by Danø et al., 1985).

The serine proteases of the plasminogen activation system have traditionally been considered as part of the haemostatic mechanism owing to the dissolution of fibrin clots by plasmin (fibrinolysis). However, the plasminogen activation system has also been implicated in cellular migration, invasion and angiogenesis (Andreasen et al., 1997; 2000; Pepper, 2001; Rakic et al., 2003). To some extent the existence of two distinct plasminogen activators accounts for these apparently distinct functions, with tissue plasminogen activator (tPA) being the primary fibrinolytic activator and urokinase-type plasminogen activator (uPA) the primary cellular activator which converts the pro-enzyme plasminogen to its active derivative, plasmin. Both plasminogen activators (PAs) are controlled by plasminogen activator inhibitors (PAIs), of which PAI-1 appears to be the predominant physiological inhibitor. Whereas tPA is primarily involved in clot dissolution (generation of plasmin for thrombolysis), uPA is recruited to the cell membrane immediately after its secretion via a specific uPA receptor (uPAR) expressed on the cell surface, and plays a role in localized cell-associated proteolysis which is an important process for cancer invasion, angiogenesis, tissue remodeling, cardiovascular complications, wound healing, and the immune response. Thus, despite their common enzymatic activities, the two PAs appear to play distinct roles in the organism. Moreover there are two main inhibitors of plasminogen activators, PAI-1 and PAI-2, while plasmin itself is inhibited by α_2-anti-plasmin (α_2AP).

The urokinase receptor (uPAR) is a cell membrane-anchored binding protein for uPA, accumulating plasminogen activation activity at cell surfaces. It has been claimed that binding of uPA to uPAR is required for its role in pericellular proteolysis because it would accelerate plasminogen activation, delay inhibition by PAI-1, regulate clearance of uPA, and localize plasmin proteolysis to the cell surface at the leading edge of the migrating cell.

Numerous studies have shown the relationship between the level of expression of uPA, uPAR, PAI-1 and their aggressive phenotypes of cancer. Based on evidence from experimental invasion and metastasis models and on expression patterns for components of the uPA system in tumours and normal tissues, it now seems beyond reasonable doubt that the uPA-mediated pathway of plasminogen activation plays a central role in tumour biology. In general, malignant tumours originating from the brain, colon, stomach, uterus, ovary, breast, kidney, colon and prostate express higher amounts or show higher activity of these enzymes than their normal or benign counterparts (Andreasen et al., 1997; 2000; Pepper, 2001; Rakic et al., 2003). Also more aggressive tumours show higher amounts of uPA system elements than less aggressive malignant tumours. Consequently, it has been established that in extensively investigated tumours such as breast cancers, the expression of uPA and PAI-1 can be used as prognostic markers predicting the outcome of the disease (Duffy et al., 1999).

However, the uPA system appears to be involved not only in cancer cell migration and invasion, but also in other tumour processes which may collectively be called "cancer cell-directed tissue remodelling". Examples of such processes are angiogenesis (Pepper, 2001c) and desmoplasia, i.e., stimulation of fibroblast proliferation and extracellular matrix protein synthesis. Although such processes involve migration and invasion by non-cancer cells, these may influence tumour growth, invasion and metastasis as well. Andreasen et al.

(2000) investigated such processes, and they have concluded that such a system has plasmin-independent functions, based on the interactions of PAI-1 and uPAR with integrins and the ECM protein vitronectin and of plasminogen activator-inhibitor complexes with endocytosis receptors. Nevertheless, this evidence was, to a large extent, obtained with non-cancer cells, but as has been stated in the previous section, no qualitative differences exist between cancer cells and normal cells (such as endothelial cells) with respect to the basic processes of migration and invasion.

In this regard, Danø et al. (1994) pointed out the importance of the stromal/cancer cell interaction in cancer invasion, and of the "orchestration" of the various proteases in this complex process. In their view, cancer mimics specific tissue remodelling processes with invasion as a form of uncontrolled tissue remodelling. Indeed, the same cell types that express specific components of the proteolytic system in remodelling processes also do so in cancer invasion. In squamous skin cancer (as in skin wound healing), the epithelial or the cancer cells, in addition to macrophages, produce both uPA and uPAR. In colon cancer, as in the shedding of epithelial cells in gastrointestinal tract, uPA is produced by fibroblast-type cells, uPAR by cancer (or epithelial) cells, and PAI-1 by endothelial cells. In mammary cancer, as in mamary gland involution, uPA is produced by myofibroblasts, while uPAR is not produced by cancer cells but mostly by macrophages. For this reason, it is important to study the remodelling processes of normal tissues in more detail.

In the following sections, we will describe the recent rapid increase in knowledge about the cell biology and regulation of plasminogen activation in relation to cell adhesion, tumour growth, cell migration, cell invasion and metastasis. Rather than being comprehensive, this section will cover specific areas related to the subsequent mathematical intrepretation of the plasminogen activation system in later chapters of this thesis. For a more detailed biochemical analysis of the model the reader is referred to the reviews by Andreasen et al. (1997), Irigoyen et al. (1999), and Andreasen et al. (2000).

3.6 Roles and Components of the PA System

Tumour expansion and dissemination can be considered as an unregulated tissue remodelling process which progressively involves both cancer and normal cells. Recruitment and reorganization of the normal host cells progressively impairs organ function, as neoplastic cells grow and influence the development of a supporting stroma infiltrated by a new blood capillary network. Tumour cell invasion and metastatic processes require the coordinated and temporal regulation of a series of adhesive, proteolytic and migratory events. Proteases have the potential to breach the mechanical barriers imposed by the basement membrane and surrounding extracellular matrix components, a prerequisite for endothelial, inflammatory or cancerous cell migration to distant sites. They have also been implicated in the activation of cytokines or other proteases, as well as in the release of growth factors sequestered within the extracellular matrix. Recent information has underlined the importance of cell-surface proteases, their receptors/activators or their inhibitors in cell migration, cell adhesion as well as in cancer cell invasion and metastasis. Novel information has recently been reported regarding the urokinase plasminogen activator system that underlines the importance of another signalling system in the mediation and regulation of cell recruitment and metastasis, although its role in extracellular matrix degradation goes back several decades (see the review by Danø et al., 1985). This section will review the available information on the pro-metastatic activity of the uPA system, its regulation and the molecular mechanisms involved.

The enzymatic system consists of the urokinase receptor (uPAR), urokinase plasminogen activator (uPA), the matrix-like protein vitronectin (VN) and plasminogen activator inhibitors: type-1 (PAI-1) and type-2 (PAI-2). uPA is an extracellular serine protease produced from cells as a single-chain proenzyme pro-uPA. Two major functional domains make up the uPA molecule: the protease domain and the growth factor domain. The protease moiety activates plasminogen and, hence, generates plasmin, a serine protease capable of digesting basement membrane and extracellular matrix proteins (see Fig. 1). Plasmin itself is a broadly acting enzyme that not only catalyzes the breakdown of many of the known extracellular matrix (ECM) and basement membrane molecules, such as vitronectin, fibrin, laminin and collagens, but also may activate metalloproteinases. Thus, the unrestrained generation of plasmin from plasminogen by the action of plasminogen activator (PA) is potentially hazardous to cells. In this regard, the process of plasminogen activation in a healthy organism is strictly controlled through the availability of PAs, localized activation, and interaction with specific inhibitors (PAIs). One of these inhibitors, PAI-1, which is believed to be the most abundant, fast-acting inhibitor of uPA in vivo (Andreasen et al., 1997; 2000). In other words, for cells to protect themselves they must secrete a surplus of inhibitors to guarantee restraint of pericellular proteolysis. Indeed secreted uPA is often associated with plasminogen activator inhibitor-1 (PAI-1) and remains inactive.

The growth factor domain has no protease activity but can bind a specific, high affinity cell-surface receptor, uPAR (or CD87). uPAR is expressed in considerable amounts on the cell surface of various cell types and as implied by the name, it was first identified as a high-affinity receptor for uPA. Additionally, uPAR mediates the binding of the zymogen pro-uPA to the plasma membrane where plasmin converts pro-uPA to the active zymogen, uPA, which in turn converts plasma membrane-associated plasminogen into plasmin. Importantly, uPA is not the only ligand for uPAR that is able to bind to the matrix-like form of vitronectin (VN) (see Fig. 1) and thus place emphasis on a non-proteolytic role for uPAR (Chapman 1997). uPAR contains a vitronectin binding site(s) distinct form the urokinase binding site. The strength of interaction between uPAR and vitronectin is not mutually exclusive; rather inactive pro-uPA, as well as active uPA promote VN binding. In addition, uPAR can also bind integrins at sites distinct from its uPA- and

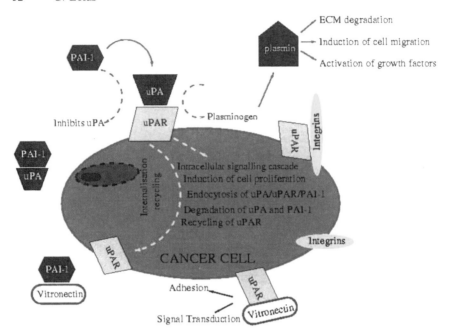

Fig. 1. Schematic diagram showing uPA binding to the cell membrane. uPA binds with high affinity to uPAR, where uPAR is anchored to the surface of a variety of cells including tumour cells. This binding activates uPA and focuses proteolytic activity to the cell surface where plasminogen is converted to plasmin. Components of the extracellular matrix such as fibronectin, laminin and vitronectin are degraded by plasmin, facilitating cell migration, angiogenesis and metastasis. Vitronectin interacts with uPAR leading to the activation of an intracellular signalling cascade

vitronectin-binding sites. These interactions account for the effects of uPAR on cell adhesion and migration.

At this point, the extracellular protein vitronectin enters the picture. Vitronectin is a versatile glycoprotein that is found in circulation, in the extracellular matrix of endothelial cells, and within various tissues of the human body (Podor et al., 2000). Vitronectin participates in the regulation of humoral responses such as coagulation and fibrinolysis. Moreover other functions of the protein that are confined to surfaces or tissues include cell-adhesion and regulation of pericellular proteolysis (Comper, 1996). As the name indicates, vitronectin binds strongly to glass surfaces (*vitro* = glass). Interactions with an assortment of biological molecules are responsible for the multiple functions exhibited by vitronectin. Defining the binding sites for these various biomolecule, along with determining the molecular mechanism of regulation, constitutes and active area of research on the protein. Work to date has focused on binding sites for several ligands, including heparin, plasminogen activator inhibitor-1, and integrins. Vitronectin binds several integrins expressed on the cell membrane, including $\alpha_v\beta_3$. It accumulates prominently in extra-cellular matrices associated with acute injury and several malignant tumours. Vitronectin functions as the major high-affinity binding protein of plasmino-gen activator inhibitor type-1 (PAI-1) and urokinase plasminogen activator receptor (uPAR). The finding that uPAR contains a high affinity binding site for vitronectin may elucidate previously unexplained and paradoxical obser-vations regarding this receptor.

First, these observations may account, at least in part, for the restricted cell surface localization of urokinase-occupied uPAR to focal contact sites in fibroblasts. These contact sites are known to co-localize with vitronectin in adherent cell lines, and the presence of vitronectin has been shown to redistribute cell surface uPAR receptors. Second, these observations raise an alternative interpretation of the paradox that increased PAI-1 is associated with enhanced cellular movement, e.g. metastasis, even though this should decrease cell surface proteolytic activity. Vitronectin is recognized as the major binding protein of PAI-1, and its binding to uPAR could be expected to bring PAI-1 in close approximation with uPA, thereby promoting inhibition and clearance of uPA from the receptor. We postulate that this process may effect a lower avidity of cellular attachment to vitronectin. In this paradigm, PAI-1, although decreasing uPA activity, would also promote detachment of the cell from its contact site. Thus, PAI-1 in circumstances where sustained proteolytic activity is not vital to movement could *promote* rather than retard migration.

PAI-1, the inhibitor of uPA, belongs to the serpin (**ser**ine **p**rotein **in**hibitors) family (Alberts et al., 1994) and can specifically bind to and in-hibit not only free, but also receptor-bound uPA (see Fig. 1). When PAI-1 is available, it can bind to the uPA/uPAR complex triggering the internal-ization of the uPA/uPAR/PAI-1 complex by receptor-mediated endocytosis. The uPA/uPAR/PAI-1 complex will be dissociated and PAI-1 and uPA will be digested, but the receptor will be recycled to the cell surface and concen-trate the uPA (if available) on the cell surface again. This process will lead to clearing of PAI-1 from the vicinity of the cell surface (Conese et al., 1995; Nykjær et al., 1997). It is not clear why PAI-1 is concentrated in the nucleus of the cancer cell. However, many receptor-binding proteins bind to the recep-tor and are then endocytosed. It is, therefore, conceivable that such protein signalling (or their degradation products) acts directly within the cell, or cell nucleus. It has been reported that the receptor-mediated internalization of uPA/uPAR/PAI-1 complexes may trigger the proliferation of the cancer cells. Additionally, inhibition of cell adhesion and migration by PAI-1 on VN occurs because the same region of VN is required for interaction with PAI-1, uPAR and integrins. In other words, PAI-1 competes with uPAR for binding to VN.

The uPA/uPAR/PAI-1/VN system therefore appears to be a very im-portant function in the regulation of the attachment/detachment machinery, namely to inform cells when, how and where to move. Availaible data suggests that cells respond to a "go" signal through the stimulation of surface prote-olysis, exposure to chemotactic epitope(s), and recycling of "naked" uPAR to novel surface proteolysis, and to a "stop" signal via PAI-1-dependent internalization and degradation of uPA. Additionally, cells respond to a "pause" signal through transient uPAR-dependent adhesion stages, thus shifting the cells between an "adhesion – mode" and a "migration – mode". Thus occupation of cell surface uPAR by uPA and concomitant urokinase activity are ephemeral in the settings of this protease inhibitor.

3.7 Proteolytic Stimulation of Cell Migration

Cell migration plays a central role in a wide variety of physiological and pathophysiological processes, for instance embryonal development, inflammation, and cancer metastasis (for reviews see Lauffenburger and Horwitz, 1996). Cell migration is the locomotion of a cell on a substratum of extracellular matrix (ECM) proteins. Cells require attachment sites on extracellular matrices in order to reorganize their cytoskeleton and initiate protrusions important to migration. In this regard, cancer cells require a well-regulated, pericellular proteolysis to migrate. They must cleave linkages to the extracellular matrix and to other cells and degrade barriers like the basement membrane, the destruction of which is a common observation in invasive cancer as well as in normal pathological situations such as wound healing.

Although proteolysis and migration through tissue barriers are normal cell functions in specific physiological circumstances, it is clear that a general aspect of malignant neoplasms includes a shift toward sustained invasive capacity. For invasion to take place, cyclic attachment to matrix components and subsequent release must occur in a directed and controlled manner. This implies that proteolysis, although enhanced in tumour cells, is still tightly regulated in a temporal and spatial fashion with respect to cell attachment. Proteolytic activity is the balance between the local concentration of activated enzymes and their endogenous inhibitors.

Cell migration proceeds through extension of the leading cytoplasmic edge, a process which among other events involves adhesion, mediated by several proteases and their extracellular matrix protein ligands. Such interactions lead to the generation of specific intracellular signals and reorganization of the cytoskeleton. The adhesions at the leading cellular edge are thought to provide guidance and traction for pulling the cell body forward. Dissociation of integrin/ligand and cell surface receptor/ligand complexes, via regulated signals delivered from the cell interior, allows retraction of the trailing edge (Lauffenburger and Horwitz, 1996).

This adhesion may not be so stringent as to prevent movement, nor too weak to provide traction. The extent of migration may thus vary with the avidity of adhesion. In addition, adhesion must be regulatable or reversible to allow detachment. Detachment from focal adhesive sites during migration is thought to occur by several mechanisms including cell surface proteolysis, alterations in integrin conformation, and bulk shedding of attachment sites.

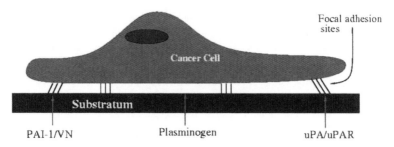

Fig. 2. Schematic diagram showing localization of uPA, uPAR, PAI-1 and plasminogen on the cell surface and in the pericellular space

Extracellular proteolytic enzyme systems like the plasminogen activation system may facilitate release of the trailing edge by degradation of extracellular matrix proteins. Several proteases are involved in cell migration and invasion, but an important role has been ascribed to the plasminogen activation system.

The effects of the plasminogen activation system in cell migration may be due to a proteolytic as well as a non-proteolytic mechanism (Andreasen et al., 1997; Tarui et al., 2001). A proteolytic mechanism of cell migration implies plasmin generation at focal adhesion sites, catalyzed by uPAR-bound uPA, which could help to break physical barriers and promote detachment of the trailing edge of the cells from matrix proteins that might impede their migration. On the other hand, with a non-proteolytic mechanism, uPA is thought to promote cell migration by enhancing adhesion at the leading edge, through stimulation of binding of uPAR to VN, modulation of uPAR/integrin interactions and/or by initiation of signal transduction cascades. It is also possible that both mechanisms operate simultaneously in migrating cells (see Fig. 2).

Soon after it was reported that uPAR contains a vitronectin-binding site, it was realized that active PAI-1 blocks the interaction of uPAR and vitronectin. In this regard, Waltz et al. (1997) and Deng et al. (1996) reported that PAI-1 regulates proteolytic activity (both on the cell surface and in solution), blocks binding of and adhesion to vitronectin by myeloid cells and also blocks binding of soluble uPAR (suPAR) – that is lacking the GPI-anchor- to vitronectin. Moreover Kanse et al. (1996) demonstrated that PAI-1 blocks binding of vitronectin to uPAR on endothelial cells. Inhibition of both uPA activity and of uPA/uPAR interactions prevents extracellular matrix degradation. On the other hand, when PAI-1 binds to vitronectin, it interferes with vitronectin recognition by integrins, thereby stimulating release of cells from the matrix and paradoxically *supporting* cell migration.

It seems possible to arrive at a model unifying the many observations by assuming that proteolytic and non-proteolytic mechanisms of uPA action on cell migration are operating simultaneously in individual migrating cells. If pro-uPA is converted to active uPA at the ventral surface of the cells, non-proteolytic mechanisms could dominate at the leading edge and proteolytic mechanisms at the trailing edge. The relative importance of the proteolytic and the non-proteolytic elements and the net effect of (pro-)uPA and PAI-1 would be expected to depend on the level of expression by the migrating cells of uPAR, endocytosis receptors, and integrins, of the composition of the ECM, of the pericellular localization of (pro-)uPA and PAI-1, of mechanisms for pro-uPA activation, and of the stimuli that induce cell motility.

3.8 Chemo- and Hapto-Taxis

Tumour cells encounter a variety of soluble and substratum-bound factors which may influence their directed migration at different stages in the process of tumour invasion and metastasis. Such factors can promote the directed movement of tumour cells by at least two mechanisms, termed chemotaxis and haptotaxis.

Chemotaxis is defined as cellular locomotion directed in response to a concentration gradient of a chemical factor in solution (Lackie and Wilkinson, 1981). Cells sense the chemical and migrate toward higher concentrationsof this substance until they reach the source secreting it. On the other hand, gradients do not have to be in solution. An adhesive molecule could be present in increasing amounts along an extracellular matrix. A cell that was constantly making and breaking adhesions with such a molecule would move from a region of low concentration to an area where that adhesive molecule was more highly concentrated. Such a phenomenon is called **haptotaxis** (Carter, 1967).

The potential importance of a chemotactic response to ECM components is apparent when considering that during the process of tumour invasion and metastasis, proteolytic degradation results in solubilization of ECM components (Aznavoorian, 1990). As a result, tumour cells could conceivably detect and respond to the soluble fragments as well as to the insoluble intact matrix molecules. Therefore, chemotaxis and haptotaxis to ECM components represent two separate and distinguishable means by which tumour cells penetrate membranes and interstitial stroma.

Chemotaxis and Signal Transduction

Induction of chemotaxis and chemokinesis by uPA has been reported in a variety of cell types (Pepper et al., 1993; Gyetko et al., 1994; Resnati et al., 1996). This activity is exerted through its specific, high affinity cell surface receptor uPAR (or CD87) (Resnati et al., 1996). This receptor is expressed by activated blood leukocytes, endothelial cells, macrophages, fibroblasts, and by different types of cells in human cancer. The receptor anchors uPA at the leading edge of migrating cells and localizes it at the focal contacts and to cell-to-cell contact sites (Blasi et al., 1987; Besser et al., 1996; Fazioli et al., 1997). These sites also contain adhesion molecules, integrins, cadherins, cytoskeleton-connecting proteins, and signal transducing molecules.

Cell motility (e.g. chemotaxis, chemokinesis, migration) stimulated by active uPA can involve plasmin generation and the subsequent degradation of ECM proteins and/or proteolytic trimming of cell surface components, including adhesion receptors and uPAR itself. uPAR has been reported to associate with many signaling molecules and to mediate signal transduction (Aguirre Ghiso et al., 1999). Much attention has been focused on the possibility that uPA binding uPAR activates intracellular signal transduction cascades. Recent reports observed that the binding of uPA to uPAR in tumour or endothelial cells activates the mitogen-activated protein kinases extracellular regulated kinase 1 and 2 (ERK-1, ERK-2) (Andreasen et al., 1997; 2000). However, a major question is how uPAR mediates cellular signaling, since uPAR is not a transmembrane molecule but belongs to the group of proteins that are tethered to the plasma membrane.

The membrane attachment of uPAR via a GPI anchor (i.e. the lack of an intra-cytoplasmic region capable of connecting with the cytoplasmic signal transducers, suggests the existence of one or more hypothetical "transmembrane adapter molecules" that connects uPAR and signalling molecules (Resnati et al., 1996). Integrins may serve as such signal transducers, and indeed uPAR has been shown to be associated in the plasma membrane with complexes of integrins and tyrosin kinases suggesting a role for these complexes in transmembrane transmission of signals via uPAR. uPA/uPAR interaction causes catalytically independent responses in endothelial cells, including chemotaxis and chemokinesis. Based on the fact that the binding of ligands to integrins initiates a signal-transduction cascade it was speculated that the reported binding of uPA to the ligand VN is involved in initiation of a signal transducing cascade (Andreasen et al., 1997; 2000).

Certain actions of uPA on different cell types in culture suggest that a signal is initiated by binding of uPA to uPAR. A chemotactic activity of uPA has long been recognized in vitro on different cell types in culture (Busso et al., 1994; Gyetko et al., 1994; Resnati et al., 1996). The chemotactic activity of uPA strictly depends on binding to its receptor uPAR: it does not occur in murine cells lacking uPAR, or containing uPAR but not recognizing human uPA; it can be restored by transfection of the uPAR; and it is inhibited by antibodies that prevent uPA/uPAR interaction (Resnati et al., 1996). Occupancy of uPAR transduces a signal that results in the movement of cells; indeed uPA binding to uPAR activates several tyrosine kinases (Busso et al., 1994; Resnati et al., 1996). It is noteworthy that the signaling pathways activated by uPA/uPAR seem to be the same pathways that induce their own expression. Thus, it is possible that over-expression of the uPA/uPAR system in tumour cells leads to a signalling loop and/or activation of additional mechanisms dependent on these molecules that contribute to enhanced pericellular proteolysis, migration and proliferation (Aguirre Ghiso et al., 1999). However, at present the details of how uPA/uPAR/integrins interact and the precise way they assemble to generate signals are still under thorough investigation.

Cell Adhesion and Haptotaxis

Both cell-cell interactions and cell-stroma interactions play an important role during the invasive cascade. Connections through cell adhesion molecules, integrins, and cadherins stabilize tissue integrity, whereas loss or alteration of these cell surface proteins has been shown to be associated with increased metastatic potential. The strength and duration of cellular interactions are modulated by (a) the repertoire of receptor expression (especially integrins); (b) the relative abundance of adhesive and counteradhesive factors in the extracellular matrix; and (c) extracellular hydrolytic enzyme systems. In this regard, pericellular proteolysis initiated by the plasminogen activator/plasmin system fulfils pivotal functions in cellular migration. Direct binding of plasminogen activators and plasminogen/plasmin to cell surface receptors or to extracellular matrix drastically increases the local concentration and the efficiency of protease formation/action. The nonclassical activities of the plasminogen activation system and the pericellular cooperation of its components with adhesion receptors, extracellular matrix (ECM) proteins and signalling molecules, have provided new insights into their role as molecular coordinators of cell adhesion.

The classical role of plasminogen activation is one counteracting cell-substratum and cell-cell adhesion, as pericellular plasmin generation leads to degradation of adhesion receptors and their extracellular matrix ligands (Mignatti and Rifkin, 1993). However, under some conditions, binding of uPA to uPAR promotes cell-substratum adhesion. In this regard, binding of uPA to uPAR stimulates the adhesion of several integrin-independent cell lines to vitronectin (VN). On the other hand, PAI-1 inhibits uPAR-dependent adhesion to vitronectin. These observations show that uPA and uPAR may also affect cell adhesion. Additionally, the avidity of uPAR for vitronectin is strongly promoted by occupancy of the receptor with uPA.

Recently, it has become clear that uPAR is involved in cell-stroma interactions and signal transducing events that are independent of its role in plasminogen activation. Both the expression pattern of uPAR and its proximity to adhesion and signalling molecules places this protease receptor at the crossroads of cellular adhesion. uPAR can be found at various locations depending on the cell type and activation state (Kjøller, 2002). For example, it can be found at the apical surface of quiescent epithelial or endothelial cells or concentrated at focal or cell-cell contacts in invasive cells in association with other proteins, such as ECM adhesion molecules, cytoskeletal elements, integrins and signalling factors. It participates in cell adhesion directly by binding to vitronectin and indirectly by modulating the affinity of integrins for their complementary ligands (Tarui et al., 2001).

uPAR has been shown to bind not only uPA, but also but also the extracellular matrix protein, vitronectin (VN). By virtue of the latter activity, it acts as an adhesion receptor. The vitronectin-binding site on uPAR is distinct from the uPA binding site. The vitronectin/uPAR complex is enhanced by the simultaneous binding of urokinase (Kanse et al., 1996). The uPA/uPAR interaction increases the binding of vitronectin to the cells meaning that, somewhat paradoxically, uPA promotes cell adhesion. In this regard, it has been proposed by Wei et al. (1994) that the interaction of uPAR binding to vitronectin takes part in a balanced attachment and release scenario, directed by the plasminogen activator inhibitor type-1, which competes with uPAR in the vitronectin binding process (Deng et al., 1996).

A further, indirect role of uPAR in adhesion is provided by interactions with certain integrins, influencing the binding properties of the latter (Tarui et al., 2001). This broadens the currently held concept of uPAR-integrin interactions, in which uPAR is proposed to interact exclusively with integrins residing on the same cell (cis interaction) as an "associated protein" that mediates signal transduction directly or through the mediation of a distinct transmembrane adaptor protein. The simultaneous recognition of vitronectin by uPAR and integrins co-localizes these two receptors to adhesion structures and directs (haptotaxis) the proteolytic activity of plasminogen systems to the matrix.

Likewise, to make matters even more complicated, active PAI-1, which is the main PA antagonist, serves as a potent competitor for vitronectin binding to uPAR and integrins and thus disrupts uPAR-mediated adhesion, but also sterically inhibits integrin binding to vitronectin. Vitronectin is considered the primary PAI-1 binding plasma protein. Like PAI-1, vitronectin is significantly increased at sites of disease, or injury, where it binds collagens, uPAR or integrins. PAI-1 also seems to play a central role in cell adhesion mediated through integrins or the uPA/uPAR complex. However, when PAI-1 inhibits uPA or when PAI-1 binds vitronectin, the uPA/uPAR complex no longer interacts with vitronectin. The higher affinity of PAI-1 to vitronectin than that of the uPA/uPAR complex to vitronectin is likely to be responsible for the release of cells from this substratum by an excess of PAI-1 (Czekay et al., 2003). Therefore Deng et al. (1996) suggested that the delicate balance between cell adhesion and cell detachment is governed by PAI-1. It is tempting to speculate that the de-adhesive effects of PAI-1 are related to the observation that high PAI-1 levels are associated with a poor prognosis for survival in several metastatic human cancers (Andreasen et al., 2000).

All studies on the PA-plasmin system cited in this chapter provide ample evidence for the requirement for this system in a large number of biological phenomena and in a variety of diseases. Cell migration, cell adhesion, angiogenesis, cancer invasion and metastasis, etc. seem too many, and the underlying mechanisms and functions too different, for one protease with a single substrate, plasminogen. The recent results on signal transduction show that uPA, in addition to its enzymatic activity, can signal through its receptor inducing a variety of activities which liken it to chemokines. The assimilation explains uPA pleiotropism and its involvement in so many different diseases.

The original idea of plasminogen activation as a rate-limiting factor in tumour invasion and metastasis has been supported by many results with in

vitro and in vivo model systems and by demonstration of correlations between patient prognosis and tumour levels. The increased knowledge about
the system has also led to the realization that the system works in a far more
complex way than described by the original hypothesis. Much evidence suggests that the system also has plasmin-independent functions, consisting in
intracellular signal transduction cascades being initiated by binding of uPA
to uPAR, in uPAR acting as a vitronectin receptor and as a regulator of integrin function, in PAI-1 acting as a regulator of uPAR and integrin binding
to vitronectin, in interactions with endocytosis receptors and in an interplay
with other proteolytic enzyme systems.

This linkage suggests to us that four molecules: uPA, PAI-1, uPAR and
vitronectin, constitute the core of an integrated dynamical system which allows spatial and temporal rearrangements of its components at cell surfaces
during cell migration and invasion. Moreover, it has become clear that the
system has a multi-functional role in tumour biology. The system seems to
function not only in cancer cell migration and invasion, but also in remodelling
of the tissue surrounding the cancer cells, which may contribute decisively to
the overall process of metastasis. As the biologies of individual tumours are
different, different processes may be rate limiting for the endpoint of metastasis in different tumours, and the importance of the uPA system may therefore
vary.

Under some conditions, the protease activity of the urokinase/plasmin
arm of this system may be more important in cellular movement and tissue
remodelling; under other conditions, the intrinsic adhesiveness of uPAR for
vitronectin, perhaps regulated by repeated uPA/PAI-1 turnover or other interactions, may be more important. In addition, the unexpected finding that
PAI-1 is a marker for a poor prognosis has been the impetus for a variety of
studies with in vitro and in vivo model systems, but they have not yet provided one unifying hypothesis for the role of PAI-1. Some observations suggest
that PAI-1 may counteract migration and invasion by inhibiting uPA, while
other observations support the hypothesis that PAI-1 is needed for the optimal function of the uPA system in these processes, by regulating cell adhesion
and by restricting proteolysis in time and space. In this regard, a dual role for
uPAR and PAI-1 has been clearly defined, with uPA playing a more ancillary
role. The counter-intuitive possibility that inhibition of PAI-1 might serve **to
inhibit** cancer invasion is one that merits further investigation.

4 The Mathematical Model of Proteolysis
and Cancer Cell Invasion of Tissue

In this section we develop a new mathematical model based on a generic solid
tumour growth, which for simplicity we assume is at the avascular stage, focusing solely on how the interactions between the cancer cells (c), urokinase
plasminogen activator (uPA) (u), plasminogen activator inhibitor-1 (PAI-1)

(p), plasmin (m) and the extracellular matrix substrate (ECM) (v) may regulate tumour invasion and metastasis. As we have described in the previous sections, plasmin is a protease which is generated at the cell surface from its inactive precursor, plasminogen, via the proteolytic activity of urokinase plasminogen activator (uPA). Plasmin is a protease, which in addition to fibrin and other proteases, cleaves many extracellular matrix proteins, including fibronectin, laminin, vitronectin and thrombospondin and can activate many of the matrix mettaloproteinases, which degrade still other matrix constituents. Plasmin also can affect the activity of cytokines and growth factors, notably TGF-beta, which influences the composition of the extracellular milieu.

In this regard, the unrestrained generation of plasminogen activator (uPA) is potentially hazardous to cells. Therefore, to maintain tissue *homeostasis* and avoid unrestrained tissue damage, plasmin activity must be tightly controlled. Such regulation is achieved at multiple "checkpoints" within the plasminogen system. A primary role in plasmin regulation is played by the availability of the plasminogen activators and their corresponding inhibitors.

We now describe the way in which the tumour cell density $c(x,t)$, the urokinase plasminogen activator (uPA) protease concentration $u(x,t)$, plasminogen activator inhibitor-1 (PAI-1) concentration $p(x,t)$, plasmin concentration $m(x,t)$ and the extracellular matrix substrate (ECM, vitronectin) $v(x,t)$ are involved in invasion and derive partial differential equations governing the evolution of each variable.

(a) **Cancer Cells:**

It is well known that pericellular proteolysis plays a crucial role in tumour cell invasion. The controlled degradation of the extracellular matrix by tumour cell-associated proteases allows tumour cells to invade surrounding tissues and gain access to the circulation. In addition, invasive cells in vivo adhere to surrounding ECM molecules via specific receptors such as integrins or urokinase plasminogen activator receptors (uPAR), and produce and secrete uPA as well as other proteases such as matrix metalloproteinases (MMPs) (Aznavoorian et al. 1992). The consequent digestion of ECM allows the cells to move into the spaces thereby created and also sets up tissue gradients, which the cells then exploit to move forwards (McCarthy et al., 1983; McCarthy et al., 1986; Taraboletti et al., 1987; Aznavoorian et al., 1990; Aznavoorian et al., 1996). Movement up concentration gradients of ECM has been reported as a mechanism enabling movement through tissues by a variety of cell types. Tumour cell motility toward high concentrations/densities of substratum-bound insolubilized components has been termed "**haptotaxis**". Additionally, the term "**chemotaxis**" has also been used to describe tumour cell motility and is referred to directed migration in a gradient of a soluble attractant (Taraboletti et al., 1987).

We assume that there is a change in cell number density due to dispersion, arising from random locomotion and we take D_c as the cell random motility coefficient, characterising how cells would disperse from higher to lower densities. In other words, this means that if the initial cancer cell profile is localized in a finite region then at all subsequent times it will be confined to a finite region whose size however could change over time. On the other hand by choosing D_c to be constant we impose that an infinitesimally small density of cancer cells penetrate the entire spatial domain immediately which is physically unrealistic. However such a choice does not affect the general framework of the process since the contribution of the chemokinetic term D_c is always the smallest in cancer cell locomotion.

The second most important term that quantifies the change in cell number density is that of the "*directional flow*" of cells due to spatial gradients of environmental stimuli, such as those stimulating chemotactic or haptotactic responses. We refer to this directed movement of tumour cells in the urokinase plasminogen activation model as chemotaxis and haptotaxis – namely a response to gradients of diffusible and non-diffusible macromolecules such as urokinase plasminogen activator (Besser et al., 1996; Fazioli et al., 1997; Blasi, 1999; Carlin et al., 2003), plasminogen activator inhibitor-1 (PAI-1) (Degryse et al., 2001; Waltz et al., 2001) and vitronectin (Aznavoorian, 1990; 1996; Bafetti et al., 1996; Kanse et al., 1996; Chapman et al., 1997; Degryse et al., 2001) respectively. To incorporate this response into our mathematical model we take the cancer cell flux (due to gradients) to be $\mathbf{J_{flux}} = \mathbf{J_{chemo}} + \mathbf{J_{hapto}}$, namely $\mathbf{J_{flux}} = \chi_c c \nabla u + \zeta_c c \nabla p + \xi_c c \nabla v$, where $\chi_c, \zeta_c, \xi_c > 0$ are the chemotactic and haptotactic coefficients respectively, characterising biased directional movement in response to spatial gradients. Additionally, we include a proliferation term. Therefore, based on the above assumptions we obtain the "word equation" below for the cancer cell density:

$$\left(\text{rate of change of cell density}\right) = \left(\text{flux due to random motion}\right)$$

$$- \left(\text{chemotaxis due to uPA}\right) - \left(\text{chemotaxis due to PAI} - 1\right)$$

$$- \left(\text{haptotaxis due to VN}\right) + \left(\text{production due to cell proliferation}\right)$$

$$+ \left(\text{proliferation due to uPA cancer cells interactions}\right)$$

Representing by D_c, χ_c, ζ_c, ξ_c, μ_2 and ϕ_{13} the random motility, uPA-mediated chemotaxis, PAI-1-mediated chemotaxis, VN-mediated haptotaxis coefficients, the cancer cell proliferation rate and the cancer cell-surface receptors recycling rate, then the above "word equation" takes the following mathematical form:

$$\frac{\partial c}{\partial x} = \underbrace{D_c \frac{\partial^2 c}{\partial x^2}}_{Random\ Motion} - \frac{\partial}{\partial x} \left(\underbrace{\chi_c\, c\, \frac{\partial u}{\partial x}}_{uPA-chemo} + \underbrace{\zeta_c\, c\, \frac{\partial p}{\partial x}}_{PAI-1-chemo} + \underbrace{\xi_c\, c\, \frac{\partial v}{\partial x}}_{VN-hapto} \right)$$

$$+ \underbrace{\phi_{13}\, c\, u}_{proliferation} + \underbrace{\mu_1\, c \left(1 - \frac{c}{c_o} \right)}_{proliferation},$$

Regarding cell proliferation, we assume that in the absence of any extracellular matrix ($v = 0$) cancer cell proliferation satisfies a logistic growth law, with μ_1 representing the proliferation rate and c_o representing the maximum sustainable cell density for cancer cells. Additionally, we include the ($\phi_{13}cu$) term representing the fact that uPA "*augments*" cell proliferation in physiological (Pleknhanova et al., 2001) and pathophysiological processes (Aguirre-Ghisso et al., 2001). It is assumed that the binding of uPA to its cell surface receptor (uPAR), forms the uPA/uPAR complex which additionally is able to bind to PAI-1 and form the resulting uPA/uPAR/PAI-1 complex (Andreasen et al., 1994; Andreasen et al., 1997; Andreasen et al., 2000). The resulting uPA/uPAR/PAI-1 complex is internalized and degraded. Consequently, uPAR is internalized as a component of the complex uPA/uPAR/PAI-1 and then coendocytosed but later re-circulated to the cell surface. The re-circulated cancer cell surface receptor gives the cell the opportunity to bind to VN and other extracellular matrix components that are also assumed to enhance cell proliferation. Additionally, another approach is that uPA bound to cancer cell-surface receptors triggers several transducing signals that promote cancer cell proliferation as well.

(b) Extracellular Matrix:

Since extracellular matrix (ECM) is "static", therefore we neglect any diffusion terms (or other "migration" terms) from its behaviour. Additionally, based on the experimental evidence that uPA activates plasminogen to produce the cancer cell-surface associated protein plasmin which in turn catalyzes the breakdown of many of the known extracellular matrix (ECM) and basement membrane molecules (such as fibronectin, laminin, vitronectin and collagen) we assume that plasmin formation either degrades the extracellular matrix upon contact or through its activation from uPA secretion (Irigoyen et al., 1999). Moreover, several studies suggest that normal cells, as well as cancer cells such as gliomas, have the ability to produce numerous ECM components (Degryse et al., 2001). On the other hand, as has previously been mentioned, the major role of PAI-1 is to inhibit uPA production and thus we assume that PAI-1 binding to uPA results indireclty in the production of VN. In other words, we assume that uPA binding to PAI-1 releases the PAI-1-bound VN and therefore gives VN the opportunity to bind to the cell-surface receptors such as uPAR and/or integrins and promote its own production through the regulation of cell-matrix-associated signal transduction pathways. Finally, we assume a logistic-type proliferation or remodelling term for the extracellular matrix.

Combining these effects we obtain the following "word equation":

$$\left(\text{Rate of change of ECM density}\right) = \left(-\text{degradation due to plasmin formation}\right)$$

$$+ \left(\text{proliferation}\right) + \left(\text{indirect growth of VN due to PAI} - 1/\text{uPA binding}\right)$$

$$- \left(\text{neutralization due to PAI} - 1\, \text{inhibition}\right).$$

Denoting by, δ the degradation rate, μ_2 the proliferation rate, ϕ_{21} the production rate of PAI-1/uPA binding, and ϕ_{22} the counterbalancing of PAI-1 binding to VN, we rewrite the "word equation" in the following form:

$$\frac{\partial v}{\partial t} = \underbrace{(-\delta\, v\, m)}_{degradation} + \underbrace{\phi_{21}\, u\, p}_{uPA/PAI-1} - \underbrace{\phi_{22}\, v\, p}_{PAI-1/VN} + \underbrace{\mu_2\, v\left(1 - \frac{v}{v_o}\right)}_{proliferation},$$

where v_o represent the maximum sustainable density for the extracellular matrix.

(c) urokinase Plasminogen Activator - uPA:

Factors influencing the urokinase plasminogen activator (uPA) concentration are assumed to be diffusion, protease production and protease decay. Specifically, uPA is secreted by the cancer cells, diffuses throughout the extracellular matrix, with constant diffusion coefficient D_u, while its binding to PAI-1 as well as to cancer cell surface receptors (uPAR) dominates its removal from the system. Combining these assumptions, yields the following "word equation":

$$\left(\text{Rate of change of uPA concentration}\right) = \left(\text{motion due to diffusion}\right)$$

$$+ \left(\text{production due to cancer cells}\right) - \left(\text{removal due to PAI} - 1\, \text{inhibition}\right)$$

$$- \left(\text{removal due to binding to cancer cells}\right).$$

If we denote by D_u, α_{31}, ϕ_{31}, and ϕ_{33} respectively, the chemoattractant's assumed diffusion coefficient, its rate of production by cancer cells, its neutralization by PAI-1 inhibition and its rate of binding to cell-surface receptors (uPAR), then the above word equation can be rewritten as:

$$\frac{\partial u}{\partial t} = \underbrace{D_u \frac{\partial^2 u}{\partial x^2}}_{Diffusion} - \underbrace{\phi_{31}\, p\, u}_{PAI-1/uPA} - \underbrace{\phi_{33}\, c\, u}_{uPA/cells} + \underbrace{\alpha_{31}\, c}_{production}. \tag{4}$$

(d) Plasminogen Activator Inhibitor − 1:

PAI-1 has a distinguished role among the factors implicated in the plasminogen activation system. Although its initial role, and the one implied by its
name, is to inhibit and prevent excessive proteolysis of the peritumoral tissue, recent findings associate PAI-1 with a much different role − namely to
promote tumour invasion and metastasis. PAI-1 is the primary inhibitor of
uPA, being produced by many cell types (Andreasen et al., 1994, Conese and
Blasi, 1995a; 1995b) while it is also present in blood plasma. In this regard,
we assume that PAI-1 inhibitor diffuses with a constant diffusion coefficient
D_p, and is either produced by the extracellular matrix as a result of uPA
binding to the cancer cells surface receptors, $u\,c$, or through the degradation
of the extracellular matrix, again as a result of uPA/uPAR regulation, $u\,c\,v$.
Moreover, it is well established that PAI-1 inhibits uPA (Andreasen et al.,
1994; 1997) and that PAI-1 in the resulting PAI-1/uPA complexes no longer
binds to VN (Andreasen et al., 1994; 1997). Regarding the aforementioned
assumptions we obtain the "word equation" below:

$$\left(\,\text{Rate of change of PAI} - 1\,\text{concentration}\,\right) \; = \; \left(\,\text{motion due to diffusion}\,\right)$$

$$+ \; \left(\,\text{production due to plasmin activation or cell secretion}\,\right)$$

$$- \; \left(\,\text{loss due to VN binding}\,\right) \; - \; \left(\,\text{loss due to uPA binding}\,\right)\,.$$

Denoting by D_p the assumed PAI-1 diffusion coefficient, α_{41} the rate of production as a result of either plasmin formation or cancer cells secretion, ϕ_{41}
the neutralization rate by uPA binding and by ϕ_{42} the neutralization rate by
VN binding, then the above word equation can be rewritten as:

$$\frac{\partial p}{\partial t} = D_p \underbrace{\frac{\partial^2 p}{\partial x^2}}_{\text{Diffusion}} - \underbrace{\phi_{41}\,p\,u}_{\text{PAI-1/uPA}} - \underbrace{\phi_{42}\,p\,v}_{\text{PAI-1/VN}} + \underbrace{(\alpha_{41}\,m)}_{\text{production}} \qquad (5)$$

(e) *Plasmin*

In examining the conservation of mass regarding plasmin concentration, we
assume that binding of uPA/uPAR provides the cell surface with a potential
proteolytic activity via activation and cell-surface co-localization of plasminogen and thus enhances rates of plasmin formation (Andreasen et al., 1997;
Chapman, 1997; Andreasen et al., 2000), while the activity of uPA, and therefore the formation of plasmin, is inhibited by the binding of the serine protease inhibitor-1 (PAI-1) to urokinase plasminogen activator (uPA) (Conese
and Blasi, 1995a; 1995b Andreasen et al., 1997; 2000). Additionally, we assume that the binding of PAI-1 to VN indirectly results in the binding of uPA
to uPAR and therefore in plasmin formation. Based on the aforementioned
assumptions we deduce the "word equation" below:

$$(\text{Rate of change of plasmin concentration}) = (\text{motion due to diffusion})$$

$$(+\text{production by cells}) - (\text{loss due to uPA/PAI} - 1 \text{ binding})$$

$$+ (\text{production due to PAI} - 1/\text{VN binding}) .$$

If we now symbolize by D_m, ϕ_{53}, ϕ_{52} and ϕ_{51}, respectively, plasmin's assumed constant diffusion coefficient, its rate of production due to uPA/uPAR binding, its rate of production due to PAI-1/VN, and its inactivation rate due to uPA inhibition by PAI-1, then we have:

$$\frac{\partial m}{\partial t} = D_m \underbrace{\frac{\partial^2 m}{\partial x^2}}_{Diffusion} - \underbrace{\phi_{51}\, p\, u}_{PAI-1/uPA} + \underbrace{\phi_{52}\, p\, v}_{PAI-1/VN} + \underbrace{\phi_{53}\, u\, c}_{uPA/cells} . \tag{6}$$

Hence the complete system of equations describing the interactions of tumour cells, ECM, uPA, PAI-1 and plasmin as described in the previous paragraphs is

$$\frac{\partial c}{\partial t} = \underbrace{D_c \frac{\partial^2 c}{\partial x^2}}_{Random\ Motion} - \frac{\partial}{\partial x} \left(\underbrace{\chi_c c \frac{\partial u}{\partial x}}_{uPA-chemo} + \underbrace{\zeta_c c \frac{\partial p}{\partial x}}_{PAI-1-chemo} + \underbrace{\xi_c c \frac{\partial v}{\partial x}}_{VN-hapto} \right)$$

$$+ \underbrace{(\phi_{13}\, c\, u)}_{proliferation} + \underbrace{\mu_1 c \left(1 - \frac{c}{c_o} \right)}_{proliferation} ,$$

$$\frac{\partial v}{\partial t} = \underbrace{(-\delta\, v\, m)\ or\ (-\delta\, u\, m\, v)}_{degradation} - \underbrace{\phi_{21}\, u\, p}_{uPA/PAI-1} + \underbrace{\phi_{22}\, v\, p}_{PAI-1/VN} + \underbrace{\mu_2 v \left(1 - \frac{v}{v_o} \right)}_{proliferation} ,$$

$$\frac{\partial u}{\partial t} = \underbrace{D_u \frac{\partial^2 u}{\partial x^2}}_{Diffusion} - \underbrace{\phi_{31}\, p\, u}_{PAI-1/uPA} - \underbrace{\phi_{33}\, c\, u}_{uPA/cells} + \underbrace{\alpha_{31}\, c}_{production} ,$$

$$\frac{\partial p}{\partial t} = \underbrace{D_p \frac{\partial^2 p}{\partial x^2}}_{Diffusion} - \underbrace{\phi_{41}\, p\, u}_{PAI-1/uPA} - \underbrace{\phi_{42}\, p\, v}_{PAI-1/VN} + \underbrace{(\alpha_{41}\, m)\ or\ (\alpha_{41}\, c)}_{production} ,$$

$$\frac{\partial m}{\partial t} = \underbrace{D_m \frac{\partial^2 m}{\partial x^2}}_{Diffusion} - \underbrace{\phi_{51}\, p\, u}_{PAI-1/uPA} + \underbrace{\phi_{52}\, p\, v}_{PAI-1/VN} + \underbrace{\phi_{53}\, u\, c}_{uPA/cells} . \tag{7}$$

We consider the system to hold on some spatial domain $\Omega = [0, 2]$ (e.g. a region of tissue) with appropriate boundary and initial conditions for each of the aforementioned variables. We assume that cancer cells, and as a consequence uPA, PAI-1 and plasmin, remain within the domain of tissue under consideration and therefore zero-flux boundary conditions are imposed on $\partial\Omega$, the boundary of Ω.

Before solving the system numerically, it is helpful to recast the problem in terms of dimensionless variables, rescaling distance with the maximum distance of the cancer cells at this early stage of invasion $L = 0.1-1$ cm, time with $\tau = \frac{L^2}{D}$ (where D represents a chemical diffusion coefficient $\sim 10^{-6}$ cm$_2$ s^{-1}, Bray, 2000), tumour cell density with c_o, ECM density with v_o, uPA concentration with u_o, PAI-1 concentration with p_o and plasmin concentration with m_o (where c_o, v_o, u_o, p_o, m_o are appropriate reference variables). The dependent variables and key parameters are rescaled thus:

$$\tilde{t} = \frac{t}{\tau}, \tilde{x} = \frac{x}{L}, \tilde{c} = \frac{c}{c_o}, \tilde{v} = \frac{v}{v_o}, \tilde{u} = \frac{u}{u_o}, \tilde{p} = \frac{p}{p_o}, \tilde{n} = \frac{m}{m_o},$$

$$\tilde{D}_c = \frac{D_c}{D}, \tilde{D}_u = \frac{D_u}{D}, \tilde{D}_p = \frac{D_p}{D}, \tilde{D}_m = \frac{D_m}{D},$$

$$\tilde{\chi}_c = \chi_c \frac{u_o}{D}, \tilde{\xi}_c = \xi_c \frac{v_o}{D}, \tilde{\zeta}_c = \zeta_c \frac{p_o}{D}, \tilde{\mu}_1 = \mu_1\tau, \tilde{\mu}_2 = \mu_2\tau,$$

$$\tilde{\delta} = \left(\delta\frac{m_o\tau}{v_o}\right) or \left(\delta\frac{m_o u_o\tau}{v_o}\right), \tilde{\alpha}_{31} = \alpha_{31}\frac{c_o}{u_o}\tau, \tilde{\alpha}_{41} = \left(\alpha_{41}\frac{m_o}{p_o}\tau\right) or \left(\alpha_{41}\frac{c_o}{p_o}\tau\right),$$

$$\tilde{\phi}_{13} = \phi_{11}u_o\tau, \tilde{\phi}_{21} = \phi_{21}\frac{u_o p_o}{v_o}\tau, \tilde{\phi}_{22} = \phi_{22}p_o\tau, \tilde{\phi}_{31} = \phi_{31}p_o\tau, \tilde{\phi}_{33} = \phi_{33}c_o\tau,$$

$$\tilde{\phi}_{41} = \phi_{41}u_o\tau, \tilde{\phi}_{42} = \phi_{42}v_o\tau, \tilde{\phi}_{51} = \phi_{51}\frac{u_o p_o}{m_o}\tau, \tilde{\phi}_{52} = \phi_{52}\frac{v_o p_o}{m_o}\tau, \tilde{\phi}_{53} = \phi_{53}\frac{u_o c_o}{m_o}\tau,$$

Dropping the tildes for notational convenience, we obtain the non-dimensional system of equations:

$$\frac{\partial c}{\partial t} = \underbrace{D_c\frac{\partial^2 c}{\partial x^2}}_{Random\ Motion} - \frac{\partial}{\partial x}\left(\underbrace{\chi_c c\frac{\partial u}{\partial x}}_{uPA-chemo} + \underbrace{\zeta_c c\frac{\partial p}{\partial x}}_{PAI-1-chemo} + \underbrace{\xi_c c\frac{\partial v}{\partial x}}_{VN-hapto}\right)$$

$$+ \underbrace{(\phi_{13}\,c\,u)}_{proliferation} + \underbrace{\mu_1\,c\,(1-c)}_{proliferation},$$

$$\frac{\partial v}{\partial t} = \underbrace{(-\delta\,v\,m)\ or\ (-\delta\,u\,m\,v)}_{degradation} - \underbrace{\phi_{21}\,u\,p}_{uPA/PAI-1} + \underbrace{\phi_{22}\,v\,p}_{PAI-1/VN} + \underbrace{\mu_2\,v\,(1-v)}_{proliferation},$$

$$\frac{\partial u}{\partial t} = \underbrace{D_u\frac{\partial^2 u}{\partial x^2}}_{Diffusion} - \underbrace{\phi_{31}\,p\,u}_{PAI-1/uPA} - \underbrace{\phi_{33}\,c\,u}_{uPA/cells} + \underbrace{\alpha_{31}\,c}_{production},\qquad(8)$$

$$\frac{\partial p}{\partial t} = \underbrace{D_p\frac{\partial^2 p}{\partial x^2}}_{Diffusion} - \underbrace{\phi_{41}\,p\,u}_{PAI-1/uPA} - \underbrace{\phi_{42}\,p\,v}_{PAI-1/VN} + \underbrace{(\alpha_{41}\,m)\ or\ (\alpha_{41}\,c)}_{production}$$

$$\frac{\partial m}{\partial t} = \underbrace{D_m\frac{\partial^2 m}{\partial x^2}}_{Diffusion} - \underbrace{\phi_{51}\,p\,u}_{PAI-1/uPA} + \underbrace{\phi_{52}\,p\,v}_{PAI-1/VN} + \underbrace{\phi_{53}\,u\,c}_{uPA/cells}.$$

In order to close the system, boundary and initial conditions, for c, v, u, p and m are required.

Boundary Conditions: Guided by the in vitro experimental protocol in which invasion takes place within an isolated system, we assume that there is no-flux of tumour cells or protease across the boundary of the domain, namely $x = 0$ and $x = 2$, in one-space dimension. These boundary conditions are represented by the following equations,

$$\left(-D_c \frac{\partial c}{\partial x} + c\chi \frac{\partial u}{\partial x} + c\zeta \frac{\partial p}{\partial x} + c\xi \frac{\partial v}{\partial x}\right) = 0, \quad \text{at } x = 0, 2, \tag{9}$$

for the cells,

$$\frac{\partial u}{\partial x} = 0, \quad \text{at } x = 0, 2, \tag{10}$$

for the urokinase plasminogen activator (uPA)

$$\frac{\partial p}{\partial x} = 0, \quad \text{at } x = 0, 2, \tag{11}$$

for the plasminogen activator inhibitor-1 (PAI-1) and

$$\frac{\partial m}{\partial x} = 0, \quad \text{at } x = 0, 2, \tag{12}$$

for plasmin.

Initial Conditions: The initial distribution of the tumour cells, the extra-cellular matrix density, the urokinase plasminogen activator (uPA), the plasminogen activator inhibitor-1 (PAI-1) concentration and the plasmin concentration are prescribed by the system of equations (13). Initially we assume that there is a cluster of cancer cells already present and that they have penetrated a short distance into the extracellular matrix while the remaining space is occupied by the matrix alone. Additionally, we assume that the uPA protease as well as the PAI-1 (inactivated) inhibitor initial concentration are proportional to the initial tumour density while we assume that the plasmin protease is not yet produced from the cancer cells. Combining the above we have the following system,

$$c(x,0) = \exp\left(\frac{-x^2}{\epsilon}\right), \quad x \in [0,2] \text{ and } \epsilon > 0,$$

$$v(x,0) = 1 - \frac{1}{2}\exp\left(\frac{-x^2}{\epsilon}\right), \quad x \in [0,2] \text{ and } \epsilon > 0,$$

$$u(x,0) = \frac{1}{2}\exp\left(\frac{-x^2}{\epsilon}\right), \quad x \in [0,2] \text{ and } \epsilon > 0, \tag{13}$$

$$p(x,0) = \frac{1}{20}\exp\left(\frac{-x^2}{\epsilon}\right), \quad x \in [0,2] \text{ and } \epsilon > 0,$$

$$m(x,0) = 0, \quad x \in [0,2]$$

where throughout the chapter we have taken ϵ to be 0.01.

4.1 Estimation of Parameters

Whenever possible parameter values are estimated from available experimental data. However, given the large number of parameters in the model to be determined, it is perhaps not surprising that several remain unquantified. Focusing on the aim of our model which is to produce certain experimentally observed events of the urokinase plasminogen activation system in a *qualitative* manner, in the cases where no experimental data could be found, parameter values were chosen to give the best qualitative numerical simulation results. This is in line with previous papers successfully simulating tumour invasion and angiogenesis (Orme and Chaplain, 1997; Byrne et al., 1998; Anderson et al., 2000). A full discussion of parameter estimation is given in Lolas (2003).

We introduce D a reference chemical diffusion coefficient e.g. $D \sim 10^{-6}$ cm^2 s^{-1} (Bray, 2000). In their model of epidermal wound healing Sherratt and Murray (1990), used values of 3×10^{-9} cm^2 s^{-1} – 5.9×10^{-11} cm^2 s^{-1} for the random motility of epidermal cells. Furthermore, in their study of individual endothelial cells (ECs), Stokes et al. (1991) calculated a random motility coefficient of $(7.1 \pm 2.7) \times 10^{-9}$ cm^2 s^{-1} for ECs migrating in a culture containing an angiogenic factor αFGF, heparin and fetal serum as well as a random motility coefficient of migrating endothelial cells with agarose overlays $(2.3 \pm 0.6) \times 10^{-9}$cm^2 s^{-1} and without agarose overlays of $(6.9 \pm 2.6) \times 10^{-9}$cm^2 s^{-1}. In agreement with the aforementioned measurements for cell dispersion Bray (2000) estimated the animal cell random motility coefficient to be $\sim 5 \times 10^{-10}$cm^2 s^{-1}. In this regard, our choice for cell dispersion will vary betweeen 10^{-9}cm^2 s^{-1} and 10^{-11}cm^2 s^{-1}, so our nondimensional value will be between: $D_c = 10^{-3} - 10^{-5}$.

For the diffusion coefficient of the chemotactic chemical, Sherratt and Murray (1990) took values of 3.1×10^{-7}cm^2 s^{-1}- 6.9×10^{-6}cm^2 s^{-1} for an activator and inhibitor chemical respectively while Chaplain et al. (1995) chose 3.3×10^{-8}cm^2 s^{-1}. Assuming that the diffusion coefficient of a diffusible chemical is in the range $10^{-6} - 10^{-9}$cm^2 s^{-1}, we obtain a dimensionless estimate of D_u, D_p in the range 0.001–1. Additionally, regarding the plasmin diffusion coefficient, Robbins et al., (1965) estimated the dimensional urokinase-activated plasmin diffusion coefficient to be 4.91×10^{-7} cm^2 s^{-1}. However, regarding the existence of uPA as well as PAI-1, we believe that the plasmin diffusion coefficient will be much smaller. Therefore we choose the dimensionless parameter D_m to be in the range of $4.91 \times 10^{-2} - 4.91 \times 10^{-4}$.

Stokes et al. (1991) estimated the chemotaxis coefficient of ECs migrating in a culture containing αFGF, to be 2600 cm^2 s^{-1} M^{-1}. Andreasen et al. (1997) estimated the blood plasma concentration of uPA to be around 20 pM while Collen et al. (1986) considered a uPA concentration around 8 nM in his studies. Choosing χ_c between 0.001–1 gives a value of u_o in the range $0.38 \times 10^{-9} M - 0.38 \times 10^{-12}$ M which is consistent with experimental measurements. In the absence of reliable empirical data, we chose the haptotaxis coefficient ξ_c to be in the range of $2.5 \times 10^{-3} - 2.5 \times 10^{-1}\,\mathrm{cm}^2\,\mathrm{s}^{-1}\,\mathrm{M}^{-1}$. Therefore, considering the fact that the vitronectin blood plasma concentration is around $4\mu\,\mathrm{M}$ (Comper, 1985) leads to a dimensionless estimate of the haptotaxis coefficient ξ_c in the range between 0.001–1.

Yu et al. (1997) estimated the doubling time of human epidermoid carminoma cells (HEp3) from in vitro proliferation experiments time to be 24 h. However, small differences in growth rate were observed but they did not bear a relation to the level of the urokinase plasminogen activator receptor (uPAR). In contrast, HEp3 with a full uPAR complement grows best when crowded. By taking the proliferation rate as the reciprocal of the cell-cycle time we get $\tilde{\mu}_1 \sim 0.042\,\mathrm{h}^{-1}$. Previously, Sherratt and Murray (1990) as well as Stokes and Lauffenburger (1991) estimated the growth rate constant to be $0.04\,\mathrm{h}^{-1}$ and $0.056\,\mathrm{h}^{-1}$ respectively, assuming that all cells are proliferating. Nevertheless, Stokes and Lauffenburger (1991), Chaplain et al. (1995), as well as Orme and Chaplain (1997), reduced the chosen value of the proliferation rate to be $0.02\,\mathrm{h}^{-1}$ in order for them to compensate with the assumption that fibronectin can inhibit endothelial cell proliferation and furthermore that during the angiogenesis process proliferation is mainly confined to a zone just proximal to the tips of the capillary sprouts. In this regard, in our numerical simulations we will choose the proliferation rate to be between $0.02\,\mathrm{h}^{-1} - 0.72\,\mathrm{h}^{-1}$, and thus obtain the dimensionless parameter of μ_1 in the range 0.05–2.

Rijken (1995) and Thorsen (1998) estimated the binding rate of uPA/PAI-1 to be around $1.8 \times 10^7\,\mathrm{M}^{-1}\,\mathrm{s}^{-1}$, Moreover Rijken (1995) estimated the plasma concentration of PAI-1 to be 0.38 nM. Additionally, several studies (Collen et al., 1986; Andreasen et al., 1997) showed that the urokinase plasminogen activator (uPA) concentration was originally described in urine and its plasma concentration levels are around 0.38 nM to 4.5 nM. However, we have to bear in mind that when PAI-1 is in excess over uPA the vast majority of cells are non-adherent and therefore are released from the VN substratum. On the other hand, when there is a plethora of uPA in the system this results in uPA binding PAI-1, therefore releasing VN and promoting the cancer cells' attachment to ECM. In this regard, without loss of generality we will assume that the reference concentrations of uPA and PAI-1, namely u_o, p_o, have more or less the same concentrations. Therefore, we obtain $\phi_{31}, \phi_{41} \sim 0.1 - 1$. The choice of $\phi_{31}, \phi_{41} \sim 0.1 - 1$, gives a dimensional parameter for the uPA/PAI-1 binding rate in the range of $\tilde{\phi}_{31}, \tilde{\phi}_{41} \sim 10^5 - 10^7\,\mathrm{M}^{-1}\,\mathrm{s}^{-1}$ which is in agreement with that obtained by Rijken (1995) and Thorsen (1988), respectively.

In addition, the blood plasma membrane concentration of VN was estimated to be $4\mu\,\mathrm{M}$ (Naski et al., 1993; Comper, 1996; Andreasen et al., 1997). However, Naski et al. (1993) as well as Waltz et al. (1997) used in their experiments a rather low concentration of vitronectin i.e. 10 nM. Additionally, Térranova et al. (1985) found that doses of fibronectin between 0.1 - 10 nM stimulated cell migration. In this regard, Orme and Chaplain (1997) as well as Chaplain and Anderson (1996) used as a reference concentration for fibronectin a value of 0.1 nM. Therefore, since we have already assumed that the reference PAI-1 concentration is more or less in the nanomolecular state then we obtain for $\phi_{22}\,\phi_{42}$ a value of $0.1-1$. The aforementioned choice of $\phi_{22}\,\phi_{42} \sim 0.1-1$ gives a dimensional parameter for the VN/PAI-1 binding rate in the range of $\tilde{\phi}_{22}, \tilde{\phi}_{42} \sim 10^5 - 10^7\,\mathrm{M}^{-1}\,\mathrm{s}^{-1}$. This is in line with the results of several studies that showed that VN tightly binds PAI-1 with $K_d = 0.3\,\mathrm{nM}$ (Lijnen and Collen, 1995; Naski et al., 1993; Waltz et al., 1997).

Regarding the secretion rates of uPA from the cells as well as the secretion of PAI-1 from cancer cells or through plasmin formation, we choose the following dimensionless values for our parameters, $\alpha_{31} = 0.1-2$ and $\alpha_{41} = 0.1-2$. For the terms $\phi_{11}, \phi_{32}, \phi_{53}$ which actually represent functions associated with the uPA binding to the cell surface receptors, such as uPAR and integrins, in other words the proliferation rate of cells due to uPA-binding to the cell surface receptors complex (as in ϕ_{13}), the neutralization rate of uPA (as in ϕ_{33}) and the production (activation) rate of plasmin (as in ϕ_{53}). In this regard, by considering the dissociation constant of uPA bound-uPAR to be in the range of $0.1-0.5\,\mathrm{nM}$ (Andreasen et al., 1997; Ellis et al., 1999; Schliom et al., 2000) and additionally taking the uPAR concentration to be in the subnanomolar range, i.e. $0.4-23\,\mathrm{nM}$ (Ellis and Danø, 1991) then we can choose ϕ_{33}, ϕ_{53} to be in the range of 0.1–1.

There were a couple of parameters in the model that we were unable to estimate. Therefore, we chose their values in order to give the best qualitative results in the simulations. Considering the uPA secretion from the cancer cells we chose the nondimensional value of α to vary between 0.05–1, whereas for the extracellular matrix degradation rate we consider δ to vary between 1–20. Furthermore, we chose the extracellular matrix remodelling rate to be three to five times higher than the cancer cells proliferation rate, and therefore we took μ_2 in the range of 0.15–2.5. Last but not least, we consider $\tau = 10^4$ seconds. However, regarding ϕ_{13} which represents the cancer cells proliferation rate due to uPA binding to the cell-surface receptors we were unable to find experimental data associated with the aforementioned parameter. Therefore, lacking experimental data for the *indirect* and through signalling regulated cancer cell proliferation rate we choose ϕ_{13} to have a value between 0.01–1.

5 Numerical Simulation Results

To compute accurate numerical solutions in one space dimension we use the NAG library subroutine D03PCF. This method uses finite difference approximations to perform a spatial discretisation of the model equations, thereby reducing them to a system of (time-dependent) ordinary differential equations which are readily integrated (this is the method of lines). The (stiff) ODE system is solved using a backward difference formula. Full details can be found in Lolas (2003). In this section we will confine ourselves to describing the results of the model for a given set of parameter values and the implications of the results as they impact on the underlying biology of cancer cell invasion of tissue. The following numerical results were obtained using the following parameter values: $D_c = 3.5 \times 10^{-4}, D_u = 2.5 \times 10^{-3}, D_p = 3.5 \times 10^{-3}, D_m = 4.91 \times 10^{-3}, \chi = 3.05 \times 10^{-2}, \xi = 2.85 \times 10^{-2}, \zeta = 3.75 \times 10^{-2}, \delta = 8.15, \alpha_{31} = 0.215, \alpha_{41} = 0.5, \phi_{11}, \phi_{31}, \phi_{41}, \phi_{51} = 0.75, \phi_{22}, \phi_{42} = 0.55, \phi_{52} = 0.11, \phi_{13}, \phi_{33} = 0.3, \phi_{53} = 0.75, \mu_1 = 0.15, \mu_2 = 0.85$ (unless specified otherwise).

In order to examine the relative importance of PAI-1 excess over uPA in the model, we first of all consider that the production of PAI-1 is higher than that of uPA, and with all other parameters as above we produce the plots given in the sequence of Figs. 3–6. We note that by $t = 1$ (~3 hours) a group of cells has built up at the leading edge of the primary tumour. By $t = 25$ (~3 days), cancer cells start to produce more uPA which in turn activates plasmin and therefore degrades the ECM. As time evolves, at $t = 35$ (~4 days) two distinct cluster of cells have formed one near the left hand boundary and the other one at the centre of the plot as a consequence of increased cancer cell proliferation. By $t = 55$ (~6.5 days) the previously central cluster of tumour cells starts to migrate driven mainly by VN-mediated haptotaxis.

Additionally, by $t = 60$ (~7 days), in Fig. 4, new clusters of cells have formed due to increased cell proliferation both due to the logistic growth rate but also due to the uPA-cancer cell surface receptor signalling cascade. By $t = 105$ (~9 days) three main groups of cancer cells have formed with regions of extracellular matrix re-establishment also observed. As time evolves, at $t = 125$ (~14.5 days) the "anarchy" that characterizes the proliferating heterogeneity of solid tumours is observed and more clusters of cancer cells are forming while others are migrating further into the domain. At $t = 150$ (~17.5 days) more cluster of cells have generated due to unrestricted cancer cell proliferation.

At $t = 165$ (~19 days), in Fig. 5, the generation of more groups of cancer cells is observed, while by $t = 185$ (~21.5 days) the cancer cells proliferative heterogeneity is even more evident and we can note several groups of cancer cells accompanied either by extracellular matrix re-distribution or by increased levels of urokinase plasminogen activator (uPA) concentration. By $t = 250$ (~29 days), six clusters of cancer cells have formed, in which others migrate to both boundaries driven by hapto- or chemo-taxis respectively. Additionally, by $t = 310$ (~36 days) we note a re-distribution of the extracellular matrix components accompanied by a decrease in the number of cancer cell clusters.

Furthermore, we note that the rapid extracellular matrix re-establishment observed at $t = 310$ (~36 days), in Fig. 6, is followed by a substantial increase of new cancer cells grouping throughout the domain. Therefore, by $t = 410$ (~47.5 days) we observe five main clusters of cancer cells, while by $t = 450$ (~52 days) we see six clusters of cells. Finally, at $t = 500$ (~58 days) we observe a rapid extracellular matrix renewal together with a rapid increase in tumour cell cluster formation.

Having examined the tumour proliferative heterogeneity we now consider several changes in the model in order for us to examine the system behaviour

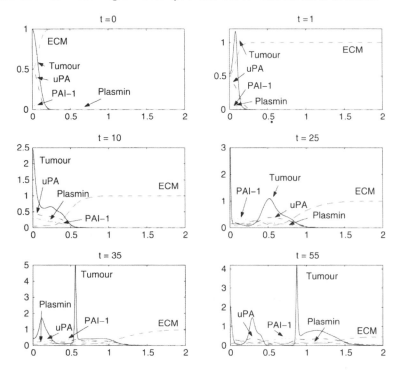

Fig. 3. Sequence of profiles showing the evolution of the tumour cell density $c(x,t)$ (*solid black line*), the uPA protease concentration $u(x,t)$ (*dot-dashed blue line*), the ECM density $v(x,t)$ (*dashed red line*), the PAI-1 concentration $p(x,t)$ (*dotted black line*) and the plasmin concentration $m(x,t)$ (*dot-dashed magenta line*). Parameter values: $D_c = 3.5 \times 10^{-4}, D_u = 2.5 \times 10^{-3}, D_p = 3.5 \times 10^{-3}, D_m = 4.91 \times 10^{-3}, \chi = 3.05 \times 10^{-2}, \xi = 2.85 \times 10^{-2}, \zeta = 3.75 \times 10^{-2}, \delta = 8.15, \alpha_{31} = 0.215, \alpha_{41} = 0.5, \phi_{21}, \phi_{31}, \phi_{41}, \phi_{51} = 0.75, \phi_{22}, \phi_{42} = 0.55, \phi_{52} = 0.11, \phi_{13}, \phi_{33} = 0.3, \phi_{53} = 0.75, \mu_1 = 0.15, \mu_2 = 0.85, L = 0.1 \, \text{cm}, \tau = 10^4$

by the inclusion or the exclusion of various other terms such as by considering cancer cell apoptosis as well as an indirect cell proliferation. Therefore, in the following set of Figs. 7 to 9 we consider a cancer cell apoptotic term. We will assume that since the uPA-binding to cancer cell surface receptors results in both having a proliferative influence over the tumour's behaviour and an activating effect in the production of plasmin, we consider that overproduction of plasmin is a signal for cancer cell apoptosis (Andreasen et al., 1997; Andreasen et al., 2000).

Additionally we assume that the PAI-1 binding to VN will result in an *indirect* cancer cell proliferation since it will eventually activate uPA-cancer-cell surface binding. Last, but not least we consider that vitronectin binding to the cancer cell surface receptors will result in a further plasmin decrease, as a result of uPA occupation by PAI-1. Therefore, the system that we will be interested in takes the following form:

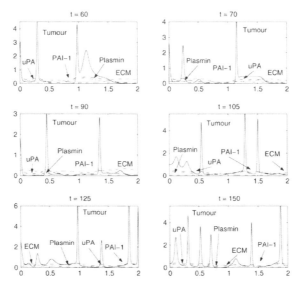

Fig. 4. Sequence of profiles showing the evolution of the tumour cell density $c(x,t)$ (*solid black line*), the uPA protease concentration $u(x,t)$ (*dot-dashed blue line*), the ECM density $v(x,t)$ (*dashed red line*), the PAI-1 concentration $p(x,t)$ (*dotted black line*) and the plasmin concentration $m(x,t)$ (*dot-dashed magenta line*). Parameter values: $D_c = 3.5 \times 10^{-4}, D_u = 2.5 \times 10^{-3}, D_p = 3.5 \times 10^{-3}, D_m = 4.91 \times 10^{-3}, \chi = 3.05 \times 10^{-2}, \xi = 2.85 \times 10^{-2}, \zeta = 3.75 \times 10^{-2}, \delta = 8.15, \alpha_{31} = 0.215, \alpha_{41} = 0.5, \phi_{21}, \phi_{31}, \phi_{41}, \phi_{51} = 0.75, \phi_{22}, \phi_{42} = 0.55, \phi_{52} = 0.55, \phi_{13}, \phi_{33} = 0.3, \phi_{53} = 0.3, \mu_1 = 0.15, \mu_2 = 0.85, L = 0.1\,\mathrm{cm}, \tau = 10^4\,\mathrm{sec}$

$$\frac{\partial c}{\partial t} = \underbrace{D_c \frac{\partial^2 c}{\partial x^2}}_{Random\ Motion} - \frac{\partial}{\partial x}\left(\underbrace{\chi_c c \frac{\partial u}{\partial x}}_{uPA-chemo} + \underbrace{\zeta_c c \frac{\partial p}{\partial x}}_{PAI-1-chemo} + \underbrace{\xi_c c \frac{\partial v}{\partial x}}_{VN-hapto} \right)$$

$$+ \underbrace{\mu_1 c(1-c) + \phi_{13} p v}_{proliferation} - \underbrace{\omega_1 m}_{apoptosis}$$

$$\frac{\partial v}{\partial t} = \underbrace{-\delta\, m\, v}_{degradation} - \underbrace{\phi_{21}\, u\, p}_{uPA/PAI-1} + \underbrace{\phi_{22}\, v\, p}_{PAI-1/VN} + \underbrace{\mu_2\, v\,(1-v)}_{proliferation},$$

$$\frac{\partial u}{\partial t} = \underbrace{D_u \frac{\partial^2 u}{\partial x^2}}_{Diffusion} - \underbrace{\phi_{31}\, p\, u}_{PAI-1/uPA} - \underbrace{\phi_{33}\, c\, u}_{uPA/uPAR} + \underbrace{\alpha_{31}\, c}_{production},$$

$$\frac{\partial p}{\partial t} = \underbrace{D_p \frac{\partial^2 p}{\partial x^2}}_{Diffusion} - \underbrace{\phi_{41}\, p\, u}_{PAI-1/uPA} - \underbrace{\phi_{42}\, p\, v}_{PAI-1/VN} + \underbrace{\alpha_{41}\, m}_{production},$$

$$\frac{\partial m}{\partial t} = \underbrace{D_m \frac{\partial^2 m}{\partial x^2}}_{Diffusion} - \underbrace{\phi_{51}\, p\, u}_{PAI-1/uPA} + \underbrace{\phi_{52}\, p\, v}_{PAI-1/VN} + \underbrace{\phi_{53}\, u\, p}_{uPA/PAI-1} - \underbrace{\phi_{54}\, c\, v}_{VN/cells}. \quad (14)$$

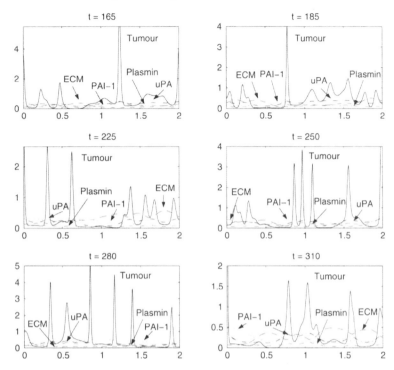

Fig. 5. Sequence of profiles showing the evolution of the tumour cell density $c(x,t)$ (*solid line*), the uPA protease concentration $u(x,t)$ (*dot-dashed line*), the ECM density $v(x,t)$ (*dashed line*), the PAI-1 concentration $v(x,t)$ (*dotted line*) and the Plasmin concentration $m(x,t)$ (*dot-dashed magenta line*). Parameter values: $D_c = 3.5 \times 10^{-4}, D_u = 2.5 \times 10^{-3}, D_p = 3.5 \times 10^{-3}, D_m = 4.9 \times 10^{-3}, \chi = 3.05 \times 10^{-2}, \xi = 2.85 \times 10^{-2}, \zeta = 3.75 \times 10^{-2}, \delta = 8.15, \alpha_{31} = 0.15, \alpha_{41} = 0.5, \phi_{13}, \phi_{33}, \phi_{53} = 0.3, \phi_{21}, \phi_{31}, \phi_{41}, \phi_{51} = 0.75, \phi_{22}, \phi_{42} = 0.55, \phi_{52} = 0.11, \mu_1 = 0.15, \mu_2 = 0.85, L = 0.1 \, \text{cm}, \tau = 10^4 \, \text{sec}$

To close our system we have to mention that we took our initial conditions to be given by the equation (13) while our boundary conditions were given by the equatons (9)–(12). The aforementioned system is solved using the following parameter values: $D_c = 10^{-4}, D_u = 2.5 \times 10^{-3}, D_p = 3.5 \times 10^{-3}, D_m = 4.9 \times 10^{-3}, \chi_c = 1.5 \times 10^{-2}, \xi_c = 2.85 \times 10^{-2}, \zeta_c = 3.5 \times 10^{-2}, \delta = 7.5, \alpha_{31} = 0.35, \alpha_{13}, \phi_{33}, \phi_{53} = 0.1, \phi_{21}, \phi_{31}, \phi_{41}, \phi_{51} = 0.15, \phi_{22} = \phi_{42} = 0.12, \phi_{52} = 0.05, \phi_{54} = 0.03, \mu_1 = 0.2, \mu_2 = 0.85, \omega_1 = 0.085$.

Based on the above assumptions, we note that at $t = 1$ (~3 hours), in Fig. 7 a large cluster of cells has built up at the leading edge of the primary tumour. However, by $t = 10$ (~1 day) due to the "taxis" competition between uPA, PAI-1-mediated and VN-mediated chemotaxis and haptotaxis respectively cancer cells are unable to migrate further into the region. As time

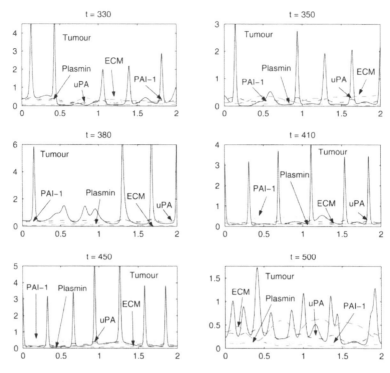

Fig. 6. Sequence of profiles showing the evolution of the tumour cell density $c(x,t)$ (*solid line*), the uPA protease concentration $u(x,t)$ (*dot-dashed line*), the ECM density $v(x,t)$ (*dashed line*), the PAI-1 concentration $p(x,t)$ (*dotted line*) and the Plasmin concentration $m(x,t)$ (*dot-dashed magenta line*). Parameter values: $D_c = 3.5 \times 10^{-4}, D_u = 2.5 \times 10^{-3}, D_p = 3.5 \times 10^{-3}, D_m = 4.9 \times 10^{-3}, \chi = 3.05 \times 10^{-2}, \xi = 2.85 \times 10^{-2}, \zeta = 3.75 \times 10^{-2}, \delta = 8.15, \alpha_{31} = 0.15, \alpha_{41} = 0.5, \phi_{13}, \phi_{33}, \phi_{53} = 0.3, \phi_{21}, \phi_{31}, \phi_{41}, \phi_{51} = 0.75, \phi_{22}, \phi_{42}, = 0.55, \phi_{52} = 0.11, \mu_1 = 0.15, \mu_2 = 0.85, L = 0.1\,\mathrm{cm}, \tau = 10^4\,\mathrm{sec}$

evolves, $t = 55$ (6 days) a large region of the extracellular matrix has been degraded as a result of uPA over-secretion by the cancer cells.

By $t = 60$ (~7 days), in Fig. 8, cancer cells start to migrate further into the region, while by $t = 70$ (~8 days) cancer cells have migrated into the degraded region. Additionally, by $t = 105$, (~12 days) more clusters of cells have formed due to the uncontrolled cell proliferation and the uPA-cancer cell surface receptors interactions. Therefore, by $t = 120$ (~14 days) we are able to observe the results of the cancer cell abnormal and irregular proliferative heterogeneity in more detail with the formation of five clusters of cancer cells. In line with previously mentioned results, we once again note the cancer cells' variability, in other words the ability of the tumour cells to produce different tumour masses following heterogeneous kinetics.

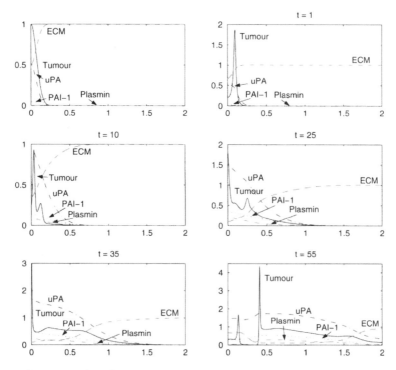

Fig. 7. Sequence of profiles showing the evolution of the tumour cell density $c(x,t)$ (*solid black line*), the uPA protease concentration $u(x,t)$ (*dot-dashed blue line*), the ECM density $v(x,t)$ (*dashed red line*), the PAI-1 concentration $v(x,t)$ (*dotted black line*) and the plasmin concentration $m(x,t)$ (*dot-dashed magenta line*). Parameter values: $D_c = 10^{-4}, D_u = 2.5 \times 10^{-3}, D_p = 3.5 \times 10^{-3}, D_m = 4.9 \times 10^{-3}, \chi_c = 1.5 \times 10^{-2}, \xi_c = 2.85 \times 10^{-2}, \zeta_c = 3.5 \times 10^{-2}, \delta = 7.5, \alpha_{31} = 0.35, \alpha_{13}, \phi_{33}, \phi_{53} = 0.1, \phi_{21}, \phi_{31}, \phi_{41}, \phi_{51} = 0.15, \phi_{22} = \phi_{42} = 0.12, \phi_{52} = 0.05, \phi_{54} = 0.03, \mu_1 = 0.2, \mu_2 = 0.85, \omega_1 = 0.085, L = 0.1\,\mathrm{cm}, \tau = 10^4$

Surprisingly, at $t = 155$ (~18 days), in Fig. 9 we point out regions of extracellular matrix redistribution which cancer cells degrade locally through the uPA production and consequently plasmin activation. Additionally, it is worth noting that these group of cells are able to migrate to either boundary and therefore provide the tumour with the ability to extravasate and migrate into distant sites where new tumours could be formed. As time evolves, $t = 250$ (29 days), the tumour heterogeneity is even more observable as a result of increased abnormal cell proliferation.

6 Discussion

In this chapter we have managed to capture the pleiotropic functions of uroki-nase plasminogen activation system related to pericellular proteolysis and

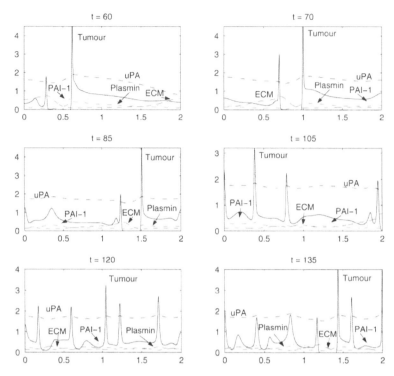

Fig. 8. Sequence of profiles showing the evolution of the tumour cell density $c(x,t)$ (*solid black line*), the uPA protease concentration $u(x,t)$ (*dot-dashed blue line*), the ECM density $v(x,t)$ (*dashed red line*), the PAI-1 concentration $v(x,t)$ (*dotted black line*) and the plasmin concentration $m(x,t)$ (*dot-dashed magenta line*). Parameter values: $D_c = 10^{-4}, D_u = 2.5 \times 10^{-3}, D_p = 3.5 \times 10^{-3}, D_m = 4.9 \times 10^{-3}, \chi_c = 1.5 \times 10^{-2}, \xi_c = 2.85 \times 10^{-2}, \zeta_c = 3.5 \times 10^{-2}, \delta = 7.5, \alpha_{31} = 0.35, \alpha_{13}, \phi_{33}, \phi_{53} = 0.1, \phi_{21}, \phi_{31}, \phi_{41}, \phi_{51} = 0.15, \phi_{22} = \phi_{42} = 0.12, \phi_{52} = 0.05, \phi_{54} = 0.03, \mu_1 = 0.2, \mu_2 = 0.85, \omega_1 = 0.085, L = 0.1 \, \text{cm}, \tau = 10^4 \, \text{sec}$

cancer cell tissue invasion. The main achievement of this chapter is that fairly simple mathematical models representing the binding interactions of the components of the plasminogen activation system coupled with cell migration were able to capture the main characteristic effects of the system in cancer progression and invasion.

We have shown in this chapter that the spatially heterogeneous distributions of cancer cell which arise as a consequence of simple binding reactions and gradient-driven migration may help to explain certain clinically and experimentally observed phenomena in carcinoma and multicellular spheroids, i.e. the heterogeneous "anarchic" spatial distribution of proliferating cancer cells and tissue.

The applications of our complete plasmin taxis-reaction-diffusion equations to cancer invasion enable us to model more realistically solid tumour

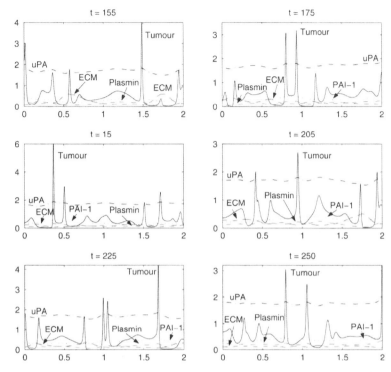

Fig. 9. Sequence of profiles showing the evolution of the tumour cell density $c(x,t)$ (*solid black line*), the uPA protease concentration $u(x,t)$ (*dot-dashed blue line*), the ECM density $v(x,t)$ (*dashed red line*), the PAI-1 concentration $v(x,t)$ (*dotted black line*) and the plasmin concentration $m(x,t)$ (*dot-dashed magenta line*). Parameter values: $D_c = 10^{-4}, D_u = 2.5 \times 10^{-3}, D_p = 3.5 \times 10^{-3}, D_m = 4.9 \times 10^{-3}, \chi_c = 1.5 \times 10^{-2}, \xi_c = 2.85 \times 10^{-2}, \zeta_c = 3.5 \times 10^{-2}, \delta = 7.5, \alpha_{31} = 0.35, \alpha_{13}, \phi_{33}, \phi_{53} = 0.1, \phi_{21}, \phi_{31}, \phi_{41}, \phi_{51} = 0.15, \phi_{22} = \phi_{42} = 0.12, \phi_{52} = 0.05, \phi_{54} = 0.03, \mu_1 = 0.2, \mu_2 = 0.85, \omega_1 = 0.085, L = 0.1\,\text{cm}, \tau = 10^4\,\text{sec}$

invasion of tissue and we believe that the results of the numerical simulations are highly consistent with in vitro as well as in vivo experimentally observed proliferative heterogeneity of cancer cells in solid tumours at their invasive stage. Especially, our models are in line with recent experimental results, that showed that when breast cells become malignant, plasmin is activated on their membrane and their morphology is changed from sheet-like structures to multicellular heterogeneous masses (Chun, 1997).

An analysis of the steady states of the model (Lolas, 2003) reveals that there no limit cycles present in the kinetic equation of our system. The dynamic,heterogeneous spatio-temporal behaviour of the system is therefore due to the interplay between proliferation, matrix degradation and taxis of the components of the uPA system. Other interesting dynamic spatio-temporal behaviour is observed for other ranges of parameters and this is explored in great depth in Lolas (2003).

Therefore, considering these reports, one may speculate that through the activation of plasmin on their membrane, micrometastatic tumour cells form multicellular clusters and thus manage to shield themselves from the action of chemotherapeutic drugs, thereby impeding chemotherapies and raising therapeutic drug doses to prohibitively high levels. In this regard, compromising the protective shield or even targeting plasmin with its α_2-antiplasmin inhibitor may be a reasonable target of new therapeutic designs for cancer therapies.

In the models described in this chapter proliferation, "taxis", invasion and signalling are strongly correlated and rely on each other for the dynamical expansion of the entire system in a heterogeneous environment. Therefore, the present findings provide more evidence that tumours are indeed complex dynamic biosystems. However, by no means can the present model claim completeness. Therefore, since invasive tumour cells are widely thought to be responsible for tumour recurrence and thus ultimately for treatment failure, the understanding of these complex mechanisms is essential in order to develop novel and more successful targeting strategies against this fatal disease.

References

1. Aguirre Ghiso J.A., Alonso D.F., Farías E.F., Gomez D.E. and and Bal De Kier Joffè E. (1999), Deregulation of the signaling pathways controlling urokinase production. Its relationship with the invasive phenotype, *European Journal of Biochemistry*, **263**, 295–304.
2. Alberts B., Bray D., Lewis J., Raff M., Roberts K., Watson J.D. (1994), *Molecular Biology of the Cell*, Garland Publishing.
3. Anderson A.R.A. and Chaplain M.A.J., Newman E.L., Steele R.J.C., and Thompson A.M. (2000), Mathematical modelling of tumour invasion and metastasis, *Journal of Theoretical Medicine*, **2**, 129–154.
4. Andreasen P.A., Sottrup-Jensen L., Kjøller L., Nykjær A., Moestrup S.K., Petersen C.M. and Gliemann J. (1994), Receptor-mediated endocytosis of plasminogen activators and activator/inhibitor complexes, *FEBS Federation of European Biochemical Societies*, **338**, 239–245.
5. Andreasen P.A., Kjøller L., Christensen L. and Duffy M.J. (1997), The urokinase-type plasminogen activator system in cancer metastasis: A review, *International Journal of Cancer*, **72**, 1–22.
6. Andreasen P.A., Egelund R. and Petersen H.H. (2000), The plasminogen activation system in tumor growth, invasion, and metastasis, *Cellular and Molecular Life Sciences*, **57**, 25–40.
7. Atkins P.W. (1989), *Physical Chemistry* 3rd edition, Oxford University Press.
8. Aznavoorian S., Stracke M.L., Krutzsch H., Schiffmann E. and Liotta L.A. (1990), Signal transduction for chemotaxis and haptotaxis by matrix molecules in tumor cells, *The Journal of Cell Biology*, **110**, 1427–1438.
9. Aznavoorian S., Murphy A.N., Stetler-Stevenson W.G. and Liotta L.A. (1993), Molecular aspects of tumor cell invasion and metastasis, *Cancer*, **71**, 1368–1383.

10. Aznavoorian S., Stracke M.L., Persons J., McClanahan J. and Liotta L.A. (1996), Integrin $\alpha_v\beta_3$ mediates chemotactic and haptotactic motility in human melanoma cells through different signaling pathways, *The Journal of Biological Chemistry*, **271**, 3247–3254.

11. Bafetti L.M., Young T.N., Itoh Y. and Stack S.M. (1998), Intact vitronectin induces matrix metalloproteinase-2 and tissue inhibitor of metalloproteinases-2 expression and enhanced cellular invasion by melanoma cells, *The Journal of Biological Chemistry*, **273**, 143–149.

12. Bajou K., Masson V., Gerard R.D., Schmitt P.M., Albert V., Praus M., Lund L.R., Frandsen T.L., Brunner N., Dano K., Fusenig N.E., Weidle U., Carmeliet G., Loskutoff D., Collen D., Carmeliet P., Foidart J.M. and Noël A. (2001), The plasminogen activator inhibitor PAI-1 controls in vivo tumor vascularization by interaction with proteases, not vitronectin: implications for antiangiogenic strategies, *The Journal of Cell Biology*, **152**, 777–784.

13. Bajpai A., and Baker J.B. (1985), Cryptic urokinase binding sites on human foreskin fibroblasts, *Biochemical and Biophysical Research Communications*, **133**, 475–482.

14. Bajzer Z., Marušić M., Vuk-Pavlović S. (1996), Conceptual frameworks for mathematical modeling of tumor growth dynamics, *Mathemtical and Computer Modelling*, **23**, 31–46.

15. Behrendt N., Ploug M., Patthy L., Houen G., Blasi F. and Danø K. (1991), The ligand-binding domain of the cell surface receptor for urokinase-type plasminogen activator, *Journal of Biological Chemistry*, **266**, 7842–7847.

16. Behrendt N., Ronne E. and Danø K. (1996), Domain interplay in the urokinase receptor. Requirement for the third domain in high affinity ligand binding and demonstration of ligand contact sites in distinct receptor domains, *Journal of Biological Chemistry*, **271**, 22885–22894.

17. Behrendt N. and Stephens R.W. (1998), The urokinase receptor, *Fibrinolysis & Proteolysis*, **12**, 191–204.

18. Behrendt N., Jensen O.N., Engelholm L.H., Mørtz E., Mann M., and Danø K. (2000), A urokinase receptor-associated protein with specific collagen binding properties, *The Journal of Biological Chemistry*, **275**, 1993–2002.

19. Behrens J., The role of cell adhesion molecules in cancer invasion and metastasis (1993), *Breast Cancer Research Treatment*, **24**, 175–184.

20. Benson D.L., Maini P.K., and Sherratt J.A. (1993), Diffusion driven instability in an inhomogeneous domain, *Bulletin of Mathematical Biology*, **55**, 365–384.

21. Besser D., Verde P., Nagamine Y., and Blasi F. (1996), Signal transduction and the u-PA/u-PAR system, *Fibrinolysis*, **10**, 215–237.

22. Blasi F., Vassalli J-D., and Danø K. (1987), Urokinase-type plasminogen activator: Proenzyme, receptor, and inhibitors, *Journal of Cell Biology*, **104**, 801–804.

23. Blasi F., Stoppelli M.P. (1999), Proteases and cancer invasion: From belief to certainty, *Biochimica et Biophysica Acta*, **1423**, R35–R44.

24. Blasi F. (1999), Proteolysis, cell adhesion, chemotaxis, and invasiveness are regulated by the u-PA-uPAR-PAI-1 system, *Thrombosis and Haemostasis*, **82**, 298–304.

25. Brattain M.G., Fine W.D., Khaled F.M., Thompson J., and Brattain D.E. (1981), Heterogeneity of malignant cells from a human colonic carcinoma, *Cancer Research*, **41**, 1751–1756.

26. Bray (2000), *Cell Movements From Molecules to Motility*, Garland Publishing.
27. Buettner H.M., Lauffenburger D.A., and Zigmond S.H. (1989), Measurement of leukocyte motility and chemotaxis parameters with the millipore filter assay, *Journal of Immunological Methods*, **123**, 25–37.
28. Busso N., Masur S.K., Lazega D., Waxman S., and Ossowski L. (1994), Induction of cell migration by pro-urokinase binding to its receptor: possible mechanism for signal-transduction in human epithelial cells, *Journal of Cell Biology*, **126**, 259–270.
29. Bussolino F., Di Renzo M.F., Ziche M., Bocchietto E., Olivero M., Naldini L., Gaudino G., Tamagnone L., Coffer A., and Comoglio P.M. (1992), Hepatocyte growth factor is a potent angiogenic factor which stimulates endothelial cell motility and growth, *Journal of Cell Biology*, **119**, 629–641.
30. Byrne H.M., Chaplain M.A.J., Pettet G.J., and McElwain D.L.S. (1998), A mathematical model of trophoblast invasion, *Journal of Theoretical Medicine*, **1**, 275–286.
31. Byrne H.M., Chaplain M.A.J., Pettet G.J. and McElwain D.L.S. (2001), An analysis of a mathematical model of trophoblast invasion, *Applied Mathematics Letters*, **14**, 1005–1010.
32. Carlin S.M., Roth M., and Black J.L. (2003), Urokinase potentiates PDGF-induced chemotaxis of human airway smooth muscle cells, *Americal Journal of Physiology*, **284**, 1020–1026.
33. Carmeliet P., and Jain R.K. (2000), Angiogenesis in cancer and other diseases, *Nature*, **407**, 249–256.
34. Carter S.B. (1967), Haptotaxis and the mechanism of cell motility, *Nature*, **213**, 256–260.
35. Chambers A.F., and Matrisian L.M. (1997), Changing views of the role of matrix metalloproteinases in metastasis, *Journal of the National Cancer Institute*, **89**, 1260–1270.
36. Chambers A.F., Groom A.C., and MacDonald I.C. (2002), Dissemination and growth of cancer cells in metastatic sites, *Nature Revies*, **2**, 563–572.
37. Chaplain M.A.J. (1995), The mathematical modelling of tumour angiogenesis and invasion, *Acta Biotheoretica*, **43**, 387–402.
38. Chaplain M.A.J., and Anderson A.R.A. (1996), Mathematical modelling, simulation and prediction of tumour-induced angiogenesis, *Invasion & Metastasis*, **16**, 222–234.
39. Chaplain M.A.J., and Byrne H.M. (1996), Mathematical modelling of wound healing and tumour growth: Two sides of the same coin, *Wounds*, 8, 42–48.
40. Chaplain M.A.J. and Orme M.E. (1998), Mathematical model of tumour-induced angiogenesis, In *Vascular Morphogenests: In Vivo, In Vitro, In Mente*, (Edited by Little C., Sage E.H., and Mironov V.), Chap. 3.4, 205–240, Birkhauser, Boston, MA.
41. Chapman H.A. (1997), Plasminogen activators, integrins, and the coordinated regulation of cell adhesion and migration, *Current Opinion in Cell Biology*, **9**, 714–724.
42. Chun M.H. (1997), Plasmin induces the formation of multicellular spheroids of breast cancer cells, *Cancer Letters*, **117**, 51–56.
43. Collen D., Zamarron C., Lijnen H.R., and Hoylaerts M. (1986), Activation of plasminogen by pro-urokinase, *Journal of Biological Chemistry*, **261**, 1259–1266.

44. Comper W.D. 1996, *Extracellular Matrix Volume 2 Molecular Components and Interactions*, Harwood Academic Publishers.

45. Conese M., and Blasi F. (1995a), The urokinase/urokinase-receptor system and cancer-invasion, *Baillière's Clinical Haematology*, **8**, 365–389.

46. Conese M., Blasi F. (1995b), Urokinase/urokinase receptor system: Internalization/degradation of urokinase-serpin complexes. Mechanism and regulation, *Biological Chemistry Hoppe-Seyler*, **376**, 143–155.

47. Czekay R-P., Aertgeerts K., Curriden S.A., and Loskutoff D.J. (2003), Plasminogen activator inhibitor-1 detaches cells from extracellular matrices by inactivating integrins, *Journal of Cell Biology*, **160**, No. 5, 781–791.

48. Damsky C.H., and Werb Z. (1992), Signal transduction by integrin receptors for extracellular matrix: cooperative processing of extracellular information, *Current Opinion in Cell Biology*, **4**, 772–781.

49. Danø K., Anderson P.A., Grondahl-Hansen J., Kristensen P., Nielsen L.S. and Skriver L. (1985), Plasminogen activators, tissue degradation and cancer, *Advanced Cancer Research*, **44**, 139–266.

50. Danø K., Behrendt N., Brunner N., Ellis V., Ploug M. and Pyke C. (1994), The urokinase receptor. Protein structure and role in plasminogen activation and cancer invasion, *Fibrinolysis*, **8** (Suppl. 1), 189–203.

51. Degryse B., Orlando S., Resnati M., Rabbani S.A., and Blasi F. (2001), Urokinase/urokinase receptor and vitronectin/$\alpha_v\beta_3$ integrin induce chemotaxis and cytoskeleton reorganization through different signaling pathways, *Oncogene*, **20**, 2032–2043.

52. Degryse B., Sier C.F.M., Resnati M., Conese M., and Blasi F. (2001), PAI-1 inhibits urokinase-induced chemotaxis by internalizing the urokinase receptor, *FEBS Letters*, **505**, 249–254.

53. Deng G., Curriden S.A., Wang S., Rosenberg S., Loskutoff D.J. (1996), Is plasminogen activator inhibitor-1 the molecular switch that governs urokinase receptor-mediated cell adhesion and release?, *Journal of Cell Biology*, **134**, 1563–1571.

54. Duffy M.J. (1993), Urokinase-type plasminogen activator and malignancy, *Fibrinolysis*, **7**, 295–302.

55. Duffy M.J., Reilly D., McDermott E., Brouillet J.P., O'Higgins N., Fennelly J.J. and Andreasen P.A. (1994), Urokinase plasminogen activator as a prognostic marker in different subgroups of patients with breast cancer, *Cancer*, **74**, 2276–2280.

56. Duffy M.J., Reilly D., O'Grady P., O'Higgins N., Fennelly J.J., and Andreasen P.A. (1995), Tissue-type and urokinase-type plasminogen activator as prognostic markers in breast cancer, *Fibrinolysis in Disease*, CRC Press New York, 14–18.

57. Duffy M.J. (1996), Proteases as prognostic markers in cancer, *Clinical Cancer Research*, **2**, 613–618.

58. Duffy M.J. (1999), Urokinase plasminogen activator: A prognostic marker in multiple types of cancer, *Journal of Surgical Oncology*, **71**, 130–135.

59. Duffy M.J. (2001), Clinical uses of tumor markers: A critical review, *Critical Reviews in Clinical Laboratory Sciences*, **38**, 225–262.

60. Edelstein-Keshet L. (1988), *Mathematical Models in Biology*, McGraw-Hill, Inc.

61. Edwards D.R., and Murphy G. (1998), Proteases-invasion and more, *Nature*, **394**, 527–528.

62. Ellis V. and Danø K. (1991), Plasminogen activation by receptor-bound uroki-nase a kinetic study with both cell-associated and isolated receptor, *Journal of Biological Chemistry*, **266**, 12752–12758.

63. Ellis V., Whawell S.A., Werner F., and Deadman J.J (1999), Assembly of urokinase receptor-mediated plasminogen activation complexes involves direct, non-active-site interactions between urokinase and plasminogen, *Biochemistry*, **38**, 651–659.

64. Estreicher A., Mühlhauser, Carpentier J-L., Orci L., and Vassali J-D. (1990), The receptor for urokinase type plasminogen activator polarizes expression of the protease to the leading edge of migrating monocytes and promotes degra-dation of enzyme inhibitor complexes, *Journal of Cell Biology*, **111**, 783–792.

65. Fazioli F. and Blasi F. (1994), Urokinase-type plasminogen activator and its receptor: new targets for anti-metastatic therapy? *Trends of Pharmacological Science*, **15**, 25–29.

66. Fazioli F., Resnati M., Sidenius N., Higashimoto Y., Appella E., and Blasi F. (1997), A urokinase-sensitive region of the human urokinase receptor is responsible for its chemotactic activity, *The EMBO Journal*, **16**, 7279–7286.

67. Fidler I.J. (1978), Tumor heterogeneity and the biology of cancer invasion and metastasis, *Cancer Research*, **38**, 2651–2660.

68. Fidler I.J. (1991), The biology of cancer metastasis or, "you cannot fix it if you do not know how it works", *Bioessays*, **13**, 551–554.

69. Fidler I.J. (2002), The pathogenesis of cancer metastasis: the 'seed and soil' hypothesis revisited, *Nature Reviews*, **3**, 1–6.

70. Fleuren G.J., Gorter A., Kuppen P.J.K., Litvinov S., and Warnaar S.O. (1995), Tumor heterogeneity and immunotherapy of cancer, *Immunological Reviews*, **145**, 91–122.

71. Folkman J. (1974), Tumour angiogenesis, *Advances in Cancer Research*, **19**, 331–358.

72. Folkman J. (1976), The vascularization of tumours, *Scientific American*, **234**, 58–73.

73. Folkman J., and Klagsbrun M. (1987), Angiogenic factors, *Science*, **235**, 442–447.

74. Ganesh S., Sier C.F.M., Vloedgiaven H.J., De Boer A., Welvaart K. (1994), Prognostic relevance of plasminogen activators and their inhibitors in colorectal cancer, *Cancer Research*, **54**, 4065–4071.

75. Gatenby R.A. (1995), Models of tumour-host interaction as competing popu-lations: Implications for tumor biology and treatment, *Journal of Theoretical Bioogy*, **176**, 447–455.

76. Gatenby R.A., and Gawlinski (1996), A reaction-diffusion model of cancer in-vasion, *Cancer Research*, **56**, 5745–5753.

77. Gatenby R.A. (1996), Application of competition theory to tumour growth: Implications for tumour biology and treatment, *European Journal of Cancer*, **32A**, 722–726.

78. Gyetko M.R., Todd R.F., III, Wilkinson C.C., and Sitrin R.G. (1994), The urokinase receptor is required for human monocyte chemotaxis *in vitro*, *Journal of Clinical Investigation*, **93**, 1380–1387.

79. Hanahan D., and Weinberg R.A. (2000), The hallmarks of cancer, *Cell*, **100**, 57–70.

80. Heppner G.H. (1984), Tumor heterogeneity, *Cancer Research*, **44**, 2259–2265.

81. Irigoyen J.P., Muñoz-Cánoves P., Montero L., Koziczak M., and Nagamine Y. (1999), The plasminogen activator system: biology and regulation, *Cellular and Molecular Life Sciences*, **56**, 104–132.
82. Jackson T.L., Lubkin S.R., Siemens S.R., Kerr N.O., Senter P.D., and Murray J.D. (1999), Mathematical and experimental analysis of localization of anti-tumour anti-body-enzyme conjugates, *British Journal of Cancer*, **80**, 1747–1753.
83. Jackson T.L., Senter P.D., and Murray J.D. (2000), Development and validation of a mathematical model to describe anti-cancer prodrug activation by antibody-enzyme conjugates, *Journal of Theoretical Medicine*, **2**, 93–111.
84. Kanse S.M., Kost C., Wilhelm O.G., Andreasen P.A., and Preissner K.T. (1996), The urokinase receptor is a major vitronectin-binding protein on endothelial cells, *Experimental Cell Research*, **224**, 344–353.
85. Karlin S. (1983), 11th R.A. Fisher Memorial Lecture – Kin Selection and Altruism, *Proceedings of the Royal Society of London*[B], **219**, 327–353.
86. Kim S.J., Shiba E., Taguchi T., Watanabe T., Tanji Y., Kimoto Y., Izukura M. and Takai S.I. (1997), Urokinase type plasminogen activator receptor is a novel prognostic factor in breast cancer, *Anticancer Research*, **17**, 1373–1378.
87. Kim S.J., and Ossowski L. (1997), Reduction in surface urokinase receptor forces malignant cells into a protracted state of dormancy, *The Journal of Cell Biology*, **137**, 767–777.
88. King R.J.B. (2000), *Cancer Biology*, Prentice Hall.
89. Kjøller L. (2002), The urokinase plasminogen activator receptor in the regulation of the actin cytoskeleton and cell motility, *Biological Chemistry*, **383**, 5–19.
90. Kohn E.C., and Liotta L.A. (1995), Molecular insights into cancer invasion: strategies for prevention and intervention, *Cancer Research*, **51**, 1856–1862.
91. Lackie and Wilkinson, *Biology of the Chemotactic Response*, Cambridge University Press.
92. Lauffenburger D.A. and Horwitz A.F. (1996), Cell migration: A physically integrated molecular process, *Cell*, **84**, 359–369.
93. Lijnen H.R. and Collen D. (1995), Mechanisms of physiological fibrinolysis, *Baillieres Clinical Haematology*, **8**, 365–389.
94. Liotta L.A., and Stetler-Stevenson W.G. (1991), Tumor invasion and metastasis as targets for cancer therapy, *Cancer Research*, **51**, 5054–5059.
95. Liotta L.A., and Clair T. (2000), Checkpoint for invasion, *Nature*, **405**, 287–288.
96. Lolas G. (2003), Mathematical Modelling of the urokinase plasminogen activation system and its role in cancer invasion of tissue, PhD thesis, University of Dundee.
97. Maini P.K., Benson D.L., and Sherratt J.A. (1992), Pattern formation in reaction diffusion models with spatially inhomogeneous diffusion coefficients, *IMA Journal of Mathematics Applied in Medicine and Biology*, **9**, 197–213.
98. Mäkinen T., Jussila L., Veikkola T., Karpanen T., Kettunen M.I., Pulkkanen K.J., Kauppinen R., Jackson D.G., Kubo H., Nishikawa S-I., Ylä-Herttuala S., and Alitalo K. (2001), Inhibition of lymphangiogenesis with resulting lymphedema in transgenic mice expressing soluble VEGF receptor-3, *Nature Medicine*, **7**, 199–205.

99. Mandriota S.J., Jussila L., Jeltsch M., Compagni A., Baetens D., Prevo R, Banerji S., Huarte J., Montesano R., Jackson D.G., Orci L., Alitalo K., Christofori G., and Pepper M.S. (2001), Vascular edothelial growth factor-C-mediated lymphangiogenesis promotes tumour metastasis, *The EMBO Journal*, **20**, 672–682.

100. Matrisian L.M. (1992), The matrix-degrading mettalloproteinases, *BioEssays*, **14**, 455–463.

101. McCarthy J.B., Palm S.L., and Furcht L.T. (1983), Migration by haptotaxis of a schwann cell tumor line to the basement membrane glycoprotein laminin, *Journal of Cell Biology*, **97**, 772–777.

102. McCarthy J.B., Hagen S.T., and Furcht L.T. (1986), Human fibronectin contains distinct adhesion- and motility-promoting domains for metastatic melanoma cells, *Journal of Cell Biology*, **102**, 179–188.

103. Mignatti P. and Rifkin D.B. (1993), Biology and biochemistry of proteinases in tumor invasion, *Physiological Reviews*, **73**, 161–195.

104. Murray J.D. (1997), The optimal scheduling of two drugs with simple resistance for a problem in cancer chemotherapy, *IMA J. Math. Appl. Med. Biol.*, **4**, 283–303.

105. Murray J.D. (2003), *Mathematical Biology: Spatial Models and Biomedical Applications*, Spinger-Verlag.

106. Naski M.C., Lawrence D.A., Mosher D.F., Podor T.J., and Ginsburg D. (1993), Kinetics of inactivation of α-thrombin by plasminogen activator inhibitor-1, *Journal of Biological Chemistry*, **268**, 12367–12372.

107. Noël A, Bajou K., Masson V., Devy L., Frankenne F., Rakic J.M., Lambert V., Carmeliet P., Foidart J.M. (1999), Regulation of cancer invasion and vascularization by plasminogen activator inhibitor-1, *Fibrinolysis & Proteolysis*, **13**, 220–225.

108. Nykjær A., Moller B., Todd III R.F., Christensen T., Andreasen P.A., Gliemann J. and Petersen C.M. (1994), Urokinase receptor. An activation antigen in human lymphocytes, *Journal of Immunology*, **152**, 505–516.

109. Nykjær A., Conese M., Christensen E.I., Olson D., Cremona O., Gliemann J. and Blasi F. (1997), Recycling of the urokinase receptor upon internalization of the uPA:serpin complexes, *The EMBO Journal*, **16**, 2610–2620.

110. Nykjær A., Kjøller L., Cohen R.L., Lawrence D.A., Gliemann J. and Andreasen P.A. (1994), Both pro-u-PA and u-PA/PAI-1 complex bind to the α2-macroglobulin receptor/LDL receptor related protein. Evidence for multiple independent contacts between the ligands and the receptor, *Annals of the New York Academy of Sciences*, **737**, 483–485.

111. Nykjær A., Møller B., Todd R.F., Christensen T., Andreasen P.A., Gliemann J. and Petersen C.M. (1994), The Urokinase Receptor: An Activation Antigen in Human T Lympocytes, *Journal of Immunology*, **152**, 505–516.

112. Okubo (1980), *Diffusion and Ecological Problems: Mathematical Models*, Springer-Verlag.

113. Orme M.E. and Chaplain M.A.J. (1996), A mathematical model of vascular tumour growth and invasion, *Mathematical and Computer Modelling*, **23**, 43–60.

114. Orme M.E., and Chaplain M.A.J. (1997), Two-dimensional models of tumour angiogenesis and anti-angiogenesis strategies, *IMA Journal of Mathematics Applied in Medicine & Biology*, **14**, 189–205.

115. Pepper M.S., Sappino A.-P., Stöcklin R., Montesano R., Orci L., and Vassalli J.-D. (1993), Upregulation of urokinase receptor expression on migrating endothelial cells, *The Journal of Cell Biology*, **122**, 673–684.

116. Pepper M.S. (2001a), Role of the matrix metalloproteinase and plasminogen activator-plasmin systems in angiogenesis, *Arteriosclerosis, Thrombosis and Vascular Biology*, **21**, 1104–1117.

117. Pepper M.S. (2001b), Extracellular proteolysis and angiogenesis, *Thrombolysis and Haemostasis*, **86**, 346–355.

118. Perumpanani A.J., Sherratt J.A., Norbury J., Byrne H.M. (1996), Biological inferences from a mathematical model for malignant invasion, *Invasion & Metastasis*, **16**, 209–221.

119. Perumpanani A.J., Simmons D.L., Gearing A.J.H., Miller K.M., Ward G., Norbury J., Schneemann M., and Sherratt J.A. (1998), Extracellular matrix-mediated chemotaxis can impede cell migration, *Proceedings of the Royal Society of London, Series B*, **265**, 2347–2352.

120. Perumpanani A.J., and Byrne H.M. (1999), Extracellular matrix concentration exerts selection pressure on invasive cells, *European Journal of Cancer*, **35**, 1274–1280.

121. Planus E., Barlovatz-Meimon G., Rogers R.A., Bonavaud S., Ingber D.E., and Wang N., Binding of urokinase to plasminogen activator inhibitor type-1 mediates cell adhesion and spreading, *Journal of Cell Science*, **110**, 1091–1098.

122. Plekhanova O., Parfyonova Y., Bibilashvily R., Domogatskii S., Stepanova V., Gulba D.C., Agrotis A., Bobik A., and Tkachuk V. (2001), Urokinase plasminogen activator augments cell proliferation and neiontima formation in injured arteries via proteolytic mechanisms, *Atherosclerosis*, **159**, 297–306.

123. Plow E.F., Freaney D.E., Plescia J., Miles L.A. (1986), The plasminogen system and cell surface: Evidence for plasminogen and urokinase receptors on the same cell type, *Journal of Cell Biology*, **103**, 2411–2420.

124. Podor T.J., Shaughnessy S.G., Blackburn M.N., and Peterson C.B. (2000), New insights into the size and stoichiometry of the plasminogen activator inhibitor type-1/vitronectin complex, *Journal of Biological Chemistry*, **275**, 25402–25410.

125. Poliakov A., Tkachuk V., Ovchinnikova T., Potapenko N., Bagryantsev S., and Stepanova V. (2001), Plasmin-dependent elimination of the growth-factor-like domain in urokinase causes its rapid cellular uptake and degradation, *Biochemical Journal*, **355**, 639–645.

126. Preissner K.T., Kanse S.M., and May A.E. (2000), Urokinase receptor: a molecular organizes in cellular communication, *Current Opinion in Cell Biology*, **12**, 621–628.

127. Rakic J.M., Maillard C., Jost, M., Bajou K., Masson V., Devy L., Lambert V., Foidart J.M., and Noël A. (2003), Role of plasminogen activator-plasmin system in tumor angiogenesis, *Cellular and Molecular Life Sciences*, **60**, 463–473.

128. Resnati M., Guttinger M., Valcamonica S., Sidenius N, Blasi F., and Fazioli F. (1996), Proteolytic cleavage of the urokinase receptor substitutes for the agonist-induced chemotactic effect, *EMBO Journal*, **9**, 467–470.

129. Rijken D.C. (1995), Plasminogen activators and plasminogen activator inhibitors: biochemical aspects, *Baillière's Clinical Haematology*, **8**, 365–389.

130. Rivero M.A., Tranquillo R.T., Buettner H.M., and Lauffenburger D.A. (1989), Transport models for chemotactic cell populations based on individual cell behaviour, *Chemical Engineering Science*, **44**, 2881–2897.

131. Robbins K.C., Summaria L., Elwyn D., and Barlow G.H. (1965), Further studies on the purification and characterization of human plasminogen and plasmin, *The Journal of Biological Chemistry*, **240**, 541–550.

132. Rodenburg K.W., Kjøller L., Petersen H.H. and Andreasen P.A. (1998), Binding of urokinase-type plasminogen activator/plasminogen activator inhibitor-1 complex to endocytosis receptors α2-macroglobulin receptor/low density lipoprotein receptor-related protein and very low density lipoprotein receptor involves basic residues in the inhibitor, *Biochemistry Journal*, **329**, 55–63.

133. Sabapathy K.T., Pepper M.S., Kiefer F., Möhle-Steinlein U., Tacchini-Cottier F., Fetka I., Breier G., Risau W., Carmeliet P., Montesano R., and Wagner E.F. (1997), Polyoma middle T-induced vascular tumor formation: The role of the plasminogen activator/plasmin system, *The Journal of Cell Biology*, **137**, 953–963.

134. Schwartz B.S., and España (1999), Two distinct urokinase-serpin interactions regulate the initiation of cell surface-associated plasminogen activation, *The Journal of Biological Chemistry*, **274**, 15278–15283.

135. Sherratt J.A., and Murray J.D. (1990), Models of epidermal wound healing, *Proceedings of the Royal Society of London. Series B*, **241**, 29–36.

136. Sherratt J.A. (1994), Chemotaxis and chemokinesis in eukaryotic cells: The Keller-Segel equations as an approximation to a detailed model, *Bulletin of Mathematical Biology*, **56**, 129–146.

137. Shliom O., Huang M., Sachais B., Kuo A., Weisel. J.W., Nagaswami C., Nassar T., Bdeir K., Hiss E., Gawlak S., Harris S., Mazar A. and Al-Roof Higazi A. (2000), Novel interactions between urokinase and its receptor, *The Journal of Biological Chemistry*, **275**, 24304–24312.

138. Stetler-Stevenson W.G., Liotta L.A., and Kleiner D.E. (1991), Extracellular matrix 6: Role of matrix metalloproteinases in tumor invasion and metastasis, *The FASEB Journal*, **7**, 1434–1441.

139. Stokes C.L., Lauffenburger D.A., and Williams S.K. (1991), Migration of individual microvessel endothelial cells: stochastic model and parameter measurement, *Journal of Cell Science*, **99**, 419–430.

140. Stokes C.L., and Lauffenburger D. (1991), Analysis of the roles of microvessel endothelial cell random motility and chemotaxis in angiogenesis, *Journal of Theoretical Biology*, **152**, 377–403.

141. Stoppelli M.P., Corti A. Soffientini A., Cassani G., Blasi F., and Assoian R.K. (1985), Differentiation-enhanced binding of the amino-terminal fragment of human urokinase plasminogen activator to a specific receptor on U937 monocytes, *Proceedings of the National Academy of Sciences USA*, **82**, 4939–4943.

142. Stott E.L., Britton N.F., Glazier J.A., and Zajac, M. (1999), Stochastic simulation of benign avascular tumour growth using the Potts model, *Mathematical and Computer Modelling*, **30**, 183–198.

143. Swanson K.R., Alvord E.C., and Murray J.D. (2000), A quantitative model for differential motility of gliomas in grey and white matter, *Cell Proliferation*, **33**, 317–329.

144. Taraboletti G., Roberts D.D., and Liotta L.A. (1987), Thrombospondin-induced tumor cell migration: Haptotaxis and chemotaxis are mediated by different molecular domains, *The Journal of Cell Biology*, **105**, 2409–2415.

145. Tarui T., Mazar A.P., Cines D.B., and Takada Y. (2001), Urokinase-type plasminogen activator receptor (CD87) is a ligand for integrins and mediates cell-cell interaction, *The Journal of Biological Chemistry*, **276**, 3983–3990.

146. Terranova V.P., DiFlorio R., Lyall R.M., Hic S., Friesel R., and Maciag T. (1985), Human endothelial cells are chemotactic to endothelial cell growth factor and heparin, *Journal of Cell Biology*, **101**, 2330–2334.

147. Thorsen S., Philips M., Semler J. (1988), Kinetics of inhibition of tissue-type and urokinase-type plasminogen activator by plasminogen activator inhibitor type 1 and type 2, *European Journal of Biochemistry*, **175**, 33–39.

148. Waltz D.A, Natkin L.R., Fujita R.M., Wei Y., and Chapman H.A. (1997), Plasmin and plasminogen activator inhibitor type 1 promote cellular motility by regulating the interaction between the urokinase receptor and vitronectin, *Journal of Clinical Investigation*, **100**, 58–67.

149. Wang Y. (2001), The role and regulation of urokinase-type plasminogen activator receptor gene expression in cancer invasion and metastasis, *Medical Research Reviews*, **21**, 146–170.

150. Wei Y., Waltz D.A., Rao N., Drummond R.J., Rosenberg S., and Chapman H.A. (1994), Identification of the urokinase receptor as an adhesion receptor for vitronectin, *The Journal of Biological Chemistry*, **269**, 32380–32388.

151. Wijnberg M.J., Slomp J., and Verheijen J.H. (1999), Immunohistochemical analysis of the plasminogen activation system and the matrix metalloproteinases in normal and atherosclerotic human vessels, *Fibrinolysis & Proteolysis*, **13**, 252–258.

152. Wyke J.A. (2000), Overview – burgeoning promise in metastasis research, *European Journal of Cancer*, **36**, 1589–1594.

Mathematical Modelling of Spatio-temporal Phenomena in Tumour Immunology

Mark Chaplain[1] and Anastasios Matzavinos[2]

[1] Division of Mathematics, University of Dundee, Dundee, DD1 4HN, Scotland
[2] School of Mathematics, University of Minnesota, Minneapolis, MN 55455, USA

1 Introduction

Cancer still remains one of the most difficult diseases to treat clinically and is one of the main causes of mortality in developed western societies. The mortality statistics for the United Kingdom for the year 2002 show that 155,180 people were registered as dying from a malignant neoplasm.[1] This figure represents 26% of all causes of death in the UK for 2002. Similar statistics hold for the United States as can be seen in Fig. 1, which shows the main causes of death in the USA during the two years 1975 and 2001.

Great effort and resources are devoted to cancer research and our understanding of cancer biology is constantly expanding. However, the overall efficiency of our current therapeutic approaches remains rather poor as Fig. 2 indicates. In particular, Fig. 2 compares the death rates associated with the two main causes of death in the USA (i.e. heart disease and cancer) during the period 1975–2001. It is evident that the reduction in the death rate associated with cancer is unfortunately considerably smaller than the one associated with heart disease.

Current patient therapies for the treatment of cancer include surgery (i.e. removal of the tumour), chemotherapy (administration of anti-cancer drugs) and radiotherapy (treatment with X-rays). Of course surgery is appropriate only for solid tumours. Although there have been great advances in patient care and treatment over the past few decades with refinement of anti-cancer drugs and medical equipment, unfortunately chemotherapy and radiotherapy both still carry major side-effects for individual patients. This is mainly due to the severe effects that these treatments have on normal, healthy proliferating cells in the patients. As a result, *the treatment* of cancers itself causes significant morbidity and mortality.

Given these facts any design of new therapeutic approaches is of great interest and one such new approach is to treat cancer using key components

---

[1]Source: Cancer Research UK.

M. Chaplain and A. Matzavinos: *Mathematical Modelling of Spatio-temporal Phenomena in Tumour Immunology*, Lect. Notes Math. **1872**, 131–183 (2006)
www.springerlink.com

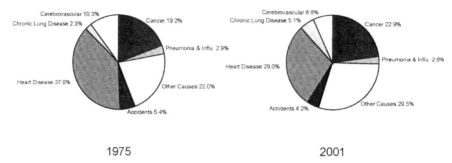

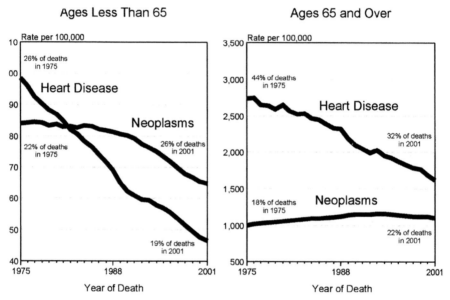

Fig. 1. Diagram showing the main causes of death in the USA in the years 1975 and 2001. Adapted from Ries et al. (2004)

Fig. 2. Distribution of death rates associated with the two main causes of death in the USA. Adapted from Ries et al. (2004)

of the immune system, the body's natural defence mechanism. In recent years there has been much biological, immunological and experimental interest in trying to develop what may be termed "immunotherapies" for cancers. One major advantage that some form of effective immunotherapy treatment would have over conventional anti-cancer treatment would be the fact that cells and other components of the immune system would be far more specific and localised in their actions, targetting cancer cells alone and leaving the vast majority of other healthy cells of the body untouched.

As part of a deeper understanding of cancer therapy the role of quantitative and predictive mathematical modelling is becoming increasingly appreciated by experimentalists and clinicians. Almost every "biological system" is by definition a highly complex, dynamic nonlinear system with many interacting variables. Because of this inherent complexity the use of modern applied mathematics has an important role to play in elucidating various aspects of the dynamical behaviour of these systems and provides a means for controlling and predicting their evolution. These statements hold in general for biological systems and also with regard to the specific topic of this chapter, the immune response to cancer. Mathematical modelling of the immune response to cancer is the main topic of interest here and we shall examine how modelling can help to shed light on a complicated process and also to be predictive and how to design the optimal immunotherapy treatments.

Therefore we start in the next section with an overview of tumour immunology. In the subsequent section we develop a continuum partial differential equation (PDE) model for the spatio-temporal response of cytotoxic T-lymphocytes to a solid tumour and we examine the spatio-temporal dynamics of the model by employing numerical simulations in a number of different biological settings. For a particular choice of parameters the model is able to simulate the phenomenon of *cancer dormancy* – a clinical condition that has been observed in breast cancers, neuroblastomas, melanomas, osteogenic sarcomas, and in several types of lymphomas. The behaviour of the cancer dormancy simulations can be described as highly irregular, depicting unstable and heterogeneous tumour cell distributions that are nonetheless characterized by a relatively low total number of tumour cells. This behaviour is consistent with several immunomorphological investigations with tumour spheroids infiltrated by cytotoxic T-lymphocytes. Finally, concluding remarks and directions for future research are given in the last section.

2 Tumour Immunology

2.1 The Immune System

The immune system is a highly complex distributed system of cells and molecules in the body that provides vertebrates with a basic defence against bacteria, viruses, fungi, and other pathogenic agents [101]. The components of the immune system can be classified as either innate or adaptive, with the latter being of interest in the framework of this chapter. Innate immune responses are stimulated by structures that are common to groups of related pathogenic agents and may not distinguish fine differences among foreign substances. The associated biological mechanisms exist before infection, are capable of rapid responses to microbes, react in essentially the same way to repeated infections and, most importantly, integrate with the adaptive components of the immune system by stimulating and influencing the nature of adaptive immune responses. Adaptive immunity develops as a response to infection and, in contrast to innate immune responses, adaptive responses are characterized by an "exquisite specificity for distinct macromolecules" and "an ability to "remember" and respond more vigorously to repeated exposures to the same microbe" [1].

The principal component of the adaptive immune system is a class of white blood cells called *lymphocytes*. There are various distinct sub-populations of lymphocytes and the cells of each sub-population express specific membrane proteins, which can be used as phenotypic markers. Two broad sub-populations are of extreme interest because they define two different types of adaptive immune responses, called humoral and cell-mediated respectively. More precisely, B-lymphocytes, which in adult mammals develop in the bone marrow, are responsible for producing antibody molecules. Pathogenic agents, also called antigens (*anti*body *gen*erators), stimulate B-lymphocytes to release antibodies into the blood, which in turn can bind to to the antigens and "mark" them as foreign structures for elimination by other cells of the immune system (e.g. macrophages). Moreover, antibodies can neutralize antigens such as viruses by coating them and preventing them from invading target cells or alternatively they can stimulate a system of blood enzymes called the *complement system*, which binds to antibody-coated structures and removes them [93]. According to [1], the term "humoral immunity" originally referred to the type of immunity that can be transferred to unimmunized individuals by antibody-containing cell-free portions of the blood, i.e. plasma or serum, which in the past were called "humors", and it is this historical framework that has established the characterization of the response of B-lymphocytes to an antigen as "humoral".

The sub-population of lymphocytes responsible for cell-mediated immune responses consists of the so-called T-lymphocytes, which develop in the thymus. T-lymphocytes express antigen receptors, which can recognize and bind to specific structures on the membrane of target cells. These structures are heterotrimers consisting of the so-called major histocompatibility complex (MHC), and a bound antigenic peptide. According to [1], the MHC was originally discovered as the genetic locus whose products were responsible for rapid rejection of tissue grafts exchanged between in-bred mice. The human MHC molecules were discovered subsequently in the context of research related to issues concerning blood transfusion and organ transplantation and in this framework they are also called human leukocyte antigens (HLAs). There are essentially two types of MHC molecules: class I and class II. Class I molecules are expressed on virtually every cell in the human body. They bind to peptides derived from proteins, which are cleaved by various proteolytic mechanisms inside the cell, and subsequently they present these peptides on the surface of the cell. Class II molecules can bind to small peptides as well, but they are expressed on a restricted set of cells, collectively characterized as professional antigen presenting cells (APCs).

The recognition of a peptide presented on the surface of a target cell as foreign by a T-lymphocyte and the associated binding of the T-lymphocyte with the target cell involve the clustering of a large number of T-lymphocyte receptors, leading to the phosphorylation of certain enzymes and triggering various cell-signaling mechanisms. Class II MHC-peptide complexes bind specifically to a sub-population of T-lymphocytes, which express the CD4 protein and hence are characterized as $CD4^+$ T-lymphocytes, also called helper T-cells. Upon antigen recognition helper T-cells essentially release substances that stimulate various components and mechanisms of both innate and adaptive immunity. Class I MHC-peptide complexes bind to $CD8^+$ T-lymphocytes, also known as cytotoxic T-lymphocytes (CTLs), which upon activation deliver apoptotic signals and kill the target cell.

Other cells of the immune system include natural killer cells, granulocytes, neutrophils, basophils, mast cells, monocytes and macrophages. All these are components of the innate immune system and will not be of main interest in the framework of this chapter. In the following sections we will focus on the response of specialized sub-populations of lymphocytes to cancer, starting in the next section with a discussion of the phenomenon of cancer dormancy in the context of tumour immunology.

2.2 Cancer Dormancy as a Manifestation of the Immune System Response to Cancer

A neoplasm (solid tumour) may be defined as "... *an abnormal mass of tissue whose growth exceeds that of normal tissue, is un-coordinated with that of the normal tissue, and persists in the same excessive manner after cessation of the stimuli which evoked the change*" [79]. A cancer, or malignant tumour, is a tumour that invades surrounding tissues, traverses at least one basement membrane zone, grows in the mesenchyme at the primary site and has the ability to grow in a distant mesenchyme, forming secondary cancers or metastases. It has been widely proposed that tumours which originate spontaneously in humans or animals often grow slowly or exist for a long period of time in a near-steady-state size even when tumour cells express activated oncogenes or enhanced growth factor signalling mechanisms. Many months, years, or even dozens of years may be required for the clinical manifestation of cancers [76, 125, 135, 140]. A solid tumour which is "near-steady-state" is described by the term *cancer dormancy* [4, 140]. The tumour nodule grows to an approximate size of 1–3 mm in diameter, containing around $10^5 - 10^6$ cells and then growth slows down and sometimes ceases. However, there are well documented clinical observations of "latent" or "dormant" human tumours containing 10^9 cells or even more [4, 15, 140]. Recent studies of the early steps in metastasis (the spread of secondary tumours) have suggested that solitary cancer cells that are neither proliferating nor undergoing apoptosis in sufficiently large numbers could contribute to metastatic recurrence after a period of "clinical dormancy" [89]. Cancer dormancy is often observed in breast cancers, neuroblastomas, melanomas, osteogenic sarcomas, and in several types of lymphomas, and is often found "accidentally" in tissue samples of healthy individuals who have died suddenly [4, 16]. In some cases, cancer dormancy has been found in cancer patients after several years of front-line therapy and clinical remission. The presence of these cancer cells in the body determines, finally, the outcome of the disease. In particular, age, stress factors, infections, act of treatment itself or other alterations in the host can provoke the initiation of uncontrolled growth of initially dormant cancer cells and subsequent waves of metastases [50, 134]. Recently, some molecular targets for the induction of cancer dormancy and the re-growth of a dormant tumour have been identified [41, 133]. However, the precise nature of the phenomenon remains poorly understood.

The early stage of primary tumour formation often occurs in the absence of a vascular network. According to [35] and [112], this stage may last up to several years. This limitation of growth is attributed by researchers to the competition between tumour cells for metabolites, a direct cytostatic/cytotoxic effect produced by the tumour cells on each other, and the competition between tumour cells and cells of the immune system for metabolites. In some cases in solid tumours there is a balance between cell proliferation and cell death. This steady-state of a fully malignant tumour (i.e. with the potential for invasion and metastases), but one which is under the local control of the host (e.g. via the immune system, endocrine system, contact inhibition) could persist for months or years [134].

One of the reasons for the slow growth of tumours and, in some cases, for their regression, may be the reaction of the host immune system to the nascent tumour cells. It has been demonstrated that tumour-associated antigens could be expressed on tumour cells at very early stages of tumour progression. Such changes are sufficient for intensive lymphoid, granulocyte and monocyte infiltration of a tumour. Especially pronounced infiltration may correlate with a favorable prognosis [17, 75, 76]. The early (avascular) stage and the subsequent stages of tumour growth are characterized by a chronic inflammatory infiltration of neutrophils, eosinophils, basophils, monocytes/macrophages, T-lymphocytes, B-lymphocytes and natural killer (NK) cells [76, 126, 141]. These cells penetrate the interior of the tumour and accumulate in it due to attractants secreted from the tumour tissue and the high locomotive ability of activated immune cells [108]. Indeed during the avascular stage, tumour development can be effectively eliminated by *tumour-infiltrating cytotoxic lymphocytes* (TICLs) [73]. The TICLs may be cytotoxic lymphocytes (CD8$^+$ CTLs), natural killer-like (NK-like) cells and/or lymphokine activated killer (LAK) cells [28, 37, 75, 141]. Cytostatic/cytotoxic activity of granulocytes and monocytes/macrophages located in the tumour is found less frequently [28, 37, 129].

An important factor, which may influence the outcome of the interactions between tumour cells and TICLs in a solid tumour, is the spatial distribution of the TICLs. A thick shell of lymphoid infiltration is often revealed around the tumour [14, 23] and even near the central hypoxic zone [74]. This would define an internal structure, whereby the regions of cell proliferation and cell death alternate, with the TICLs located near the groups of dying tumour cells [90].

In spite of some progress into the investigation of TICLs and their mechanisms of interaction with tumour cells, our understanding of the spatio-temporal dynamics of TICLs in avascular tumours and in micrometastases in vivo is still rather limited. It is perhaps not surprising therefore, that this complicated picture has not yet received an adequate explanation. Certainly, other components of the immune system (e.g. cytokines) are involved in modulating the local cellular immune response dynamics. Production of several interleukins (IL-2, IL-10, IL-12) cell-adhesion molecules (e.g. ICAM-1) and chemokines (e.g. LEC) in tumour tissue induce chemotaxis of T-cells and cytotoxic reactions of TICLs against tumour cells [21, 38, 42]. Many cytokines are produced during cell-cell interactions, which can be focussed to perform their function over short ranges in space and over short intervals of time. Strong local immune reactions are induced by the release of many interleukins, granulocyte colony-stimulating factor (G-CSF), interferons, and tumour necrosis factors. These cytokines are known to recruit and activate a variety of cell types (often in different ways), which could be tumour-infiltrating cells, or the tumour cells themselves [37, 64, 107, 121]. Besides effector immune reactions, other processes (e.g. cell proliferation, development, locomotion and apoptosis) are governed in a feedback fashion by their own intensity.

Over the last 20 years three-dimensional tissue cultures have been increasingly used to model the heterogeneity of micro-environmental and population changes which develop in solid tumours. There are two geometrically different experimental models – multicellular tumour spheroids [128] and multi-layered cell tissue [26]. Both of these experimental systems mimic the tumour environment better than mono-layer cell cultures because their spatial structure permits more cell-cell interactions and also permits the modelling of physical constraints. Numerous studies have been undertaken to examine the different mechanisms of migration and infiltration of immune cells and their interactions with the tumour cell populations within such tumour models [38, 53, 76]. However the effect of the immune cells on the tumour cells has been mainly evaluated only crudely by the survival of the tumours without a detailed spatio-temporal analysis of processes in the tumour. One obstacle to the interpretation and analysis of data obtained from such experiments is a lack of a quantitative methodology for the characterisation both of the spatio-temporal patterns of the distributions in these experimental systems, and also of the large variations at the cell level between different tumour cell types and different populations of immune cells (these differences remarkably affect the efficiency of the immune control of tumour growth).

It is difficult to control experimentally all of the interacting elements in a tumour. Furthermore, complex biological systems, such as the immune system and a cancer in vivo, do not always behave or act as predicted by experimental investigations in vitro [104]. In this framework, mathematical modelling and computer simulations can be helpful in understanding some important features associated with these elaborate systems [2, 3]. In recent years several papers have begun to investigate the mathematical modelling of various aspects of the immune system response to cancer. Specific aspects include the development of tumour heterogeneities as a result of tumour cell and macrophage interactions [96–98], macrophage infiltration into avascular tumours [58], receptor-ligand (Fas-FasL) dynamics [139], tumour progression and immune competition [5, 6, 11–13], and the dynamic of tumour cell-TICL interactions [22, 63, 65, 82, 83].

2.3 Tumour Immunology in Retrospect

According to [100] the origins of modern tumour immunology can be traced back to the 1950s, when several groups of investigators demonstrated that (a) the immune system of inbred mice can recognize antigens expressed by tumour cells induced by chemical carcinogens, (b) such recognition results in rejection of a subsequent challenge of the same tumour in previously immunized animals and (c) immune cells but not antibodies can mediate this reaction [8, 60, 95, 105]. In particular, [105] were the first to provide a definitive demonstration that chemically induced fibrosarcomas in in-bred mice express antigens that (a) elicit a transplantation immunity against the tumour and (b) are not expressed by normal cells. At the same time the concept of *immune surveillance* was being proposed by Macfarlane Burnet, stating that a physiological function of the immune system is to recognize and destroy clones of transformed cells before they grow into tumours and to kill tumours after they are formed [1, 19].

The experiments undertaken in these early studies were essentially in vivo transplantation investigations carried out on mouse models. The basic experimental procedure employed, various modifications of which were studied by [8, 60, 95] and [105], is the following. A sarcoma is induced in an in-bred mouse by painting its skin with the chemical carcinogen methylcholanthrene (MCA). If the MCA-induced tumour is surgically excised and transplanted into other syngeneic mice[2], the tumour grows. In contrast, if the tumour is transplanted back into the original host, the mouse rejects it. Moreover, one can verify that T-cells from the tumour-bearing animal can transfer protective immunity against the tumour to another tumour-free animal.

A better understanding of these experimental observations came with the development of in vitro systems that can measure in quantitative terms the cytotoxic and proliferative activity of lymphocytes against tumour cells. One of the first investigations in this direction was [18], where the authors evaluated the lytic activity of immune lymphocytes by labelling target tumour cells with the isotope chromium 51 and lymphocytes with tritiated thymidine. Furthermore, the first explorations of the reactions of immune cells against different types of human tumours were undertaken in [45]. Following the discovery of MHC restriction by Doherty and Zinkernagel in 1974 (see [29]), it was soon realized that tumour antigens can also be recognized by T-cells in an MHC-restricted fashion as demonstrated in [132]. The rapid development of basic

---

[2]Syngeneic mice have been inbred so that their chromosomes are identical.

immunology, thanks to the introduction of molecular techniques, subsequently made it possible to define the mechanisms by which antigens are presented as peptides to T-lymphocytes.

2.4 Tumour Antigens

The study of tumour antigens is a central theme in the research area of tumour immunology mainly for two reasons: (a) it reveals the molecular base of the interactions of a tumour with the tumour-bearing host's immune system and (b) as we will see in the next section, it delimits strategies for designing protocols for a broad class of therapeutic approaches called immunotherapies. The early studies in the direction of identifying tumour antigens aimed at the isolation of proteins, which are expressed exclusively by tumour cells and thus can serve as molecular markers characterizing these cells. However, it was soon realized that the vast majority of tumour antigens recognized by $CD8^+$ lymphocytes are non-mutated gene products expressed by normal cells as well. This surprising discovery led to the first classification of tumour antigens as either tumour-specific (TSAs) or tumour-associated (TAAs).

It is a remarkable aspect of tumour immunology that self-proteins can be recognized as foreign by the host's immune system when they are expressed by cancer cells. Various reasons can contribute to this phenomenon. For instance, a variety of proteins can be produced in exceptionally low quantities by normal cells failing to be recognized by the immune system and to induce tolerance. The expression of these proteins may be greatly enhanced in mutated cancer cells leading to an inflammatory reaction. Moreover, in many cases, normal proteins are produced by cells in immune privileged sites, where T-cells do not respond effectively and antigens are usually ignored. Transformed cells emanating from immune-privileged sites and expressing ignored antigens can metastasize to normal tissues, where the antigens are recognized as foreign by the host's immune system.

A variety of methods can be employed to identify tumour antigen gene products. According to [113], the majority of tumour antigens have been isolated using a genetic approach, which initially involves the generation of a cDNA library from tumour cell mRNA in a eukaryotic expression vector – that is a molecular construct that will allow expression of the genes when they are introduced into eukaryotic cell lines. Pools of cDNAs, generally containing between 100 and 200 individual cDNA clones, are produced and introduced into highly transfectable cell lines expressing the appropriate class I MHC gene product. Transfected cells are then assayed for their ability to stimulate cytokine release from tumour-reactive T-cells [113]. Alternatively, a direct biochemical approach can be employed in which peptides bound to tumour cell class I MHC molecules are eluted and fractionated according to appropriate biochemical methodologies. The antigens are identified by testing the fractions for their ability to sensitize MHC-matched non-tumour target cells for lysis by a tumour-specific CTL clone [1].

The modern classification of tumour antigens is based on their molecular structure and source of origin. It is worth mentioning that this type of classification is not uniformly consistent in the literature. For instance, according to certain authors the so-called tissue-specific differentiation antigens include various abberantly expressed normal cellular proteins (e.g. [113]), whereas according to others the definition of the notion of a tissue-specific differentiation antigen is made in such a way as to exclude abberantly expressed normal cellular proteins (e.g. [1]). Nonetheless, these differences are not of an essential nature and the common theme underlying the modern classification of tumour antigens is the understanding of their function and role in the recognition of tumour cells by the various cell types of the immune system. In what follows we present such a classification, following mainly [1] and [27].

Products of Mutated Oncogenes and Tumour Suppressor Genes

Many tumours express genes whose products are required for malignant transformation or for maintenance of the malignant phenotype. Often, these genes are produced by point mutations, deletions, chromosomal translocations, or viral gene insertions involving cellular proto-oncogenes or tumour suppresor genes to form oncogenes whose products have transforming activities [1]. Because these altered genes are not present in normal cells, they do not induce self-tolerance and peptides derived from them may stimulate T-cell responses in the host. For instance, some cancer patients have circulating $CD4^+$ and $CD8^+$ T-cells that can respond to the products of mutated oncogenes such as Ras, p53, and Bcr-Abl proteins. Furthermore, in animals, immunization with mutated Ras or p53 proteins induces CTL responses against tumours expressing these mutants. However, these proteins do not appear to be major targets of tumour-specific CTLs in most patients with a variety of tumours.

Products of Mutated Genes not Related to the Malignant Phenotype

An interesting finding from the early experiments with MCA-induced sarcomas in mouse experimental models is that different sarcomas of the same cell type, induced by the same chemical carcinogen on the same mouse are interacting with the host's immune system in different ways. The immunity that the host develops in the short time period between the excision of a sarcoma and its transplantation back into the host is specific for this very particular sarcoma. The antigens that were connected with this phenomenon were named *tumour-specific transplantation antigens* (TSTAs). We now know that they are extremely diverse mutants of cellular proteins with no relation to the malignant phenotype and no known function [1]. Their diversity can be explained by the fact that the carcinogens that induce the tumours in these experiments may randomly mutagenize virtually any host gene and the class I MHC antigen presenting pathway can display peptides from any mutated cytosolic protein in each tumour. The general principle that mutated host proteins can function as tumour antigens has been demonstrated in human cancers as well. It is however restricted to a small number of types of human cancers [1].

Aberrantly Expressed Normal Cellular Proteins

Tumour antigens may be normal cellular proteins that are over-expressed or aberrantly expressed in tumour cells. A number of CTLs raised against autologous tumour cells have been found to recognize antigens encoded by non-mutated genes expressed in both normal and tumoural tissues, although the CTLs appeared to lyse the tumour cells specifically [27, 51, 114]. For instance, HER-2/neu, which is found at high levels in about 30% of breast and ovarian carcinomas [27, 48], is expressed ubiquitously at low levels in various normal tissues. A peptide derived from HER-2/neu has been found to be the target of lymphocytes infiltrating some HLA-A2 ovarian cancers [34, 78].

Tumour Antigens Encoded by Genomes of Oncogenic Viruses

Oncogenic viruses are viruses that drive their host cell into uncontrolled proliferation and therefore can be associated with the etiology of some cancers. Examples in humans include the Epstein-Barr virus (EBV), which is associated with B-cell lymphomas and nasopharyngeal carcinoma, as well as human papilloma virus (HPV), which is the etiologic agent of cervical cancer [77, 111, 137]. Virus encoded proteins expressed by transformed cells often function as tumour antigens and elicit specific T-cell responses. Since viral peptides are foreign to the host's immune system, virus-induced tumours are among the most immunogenic tumours known. Detailed studies of antigens encoded by genomes of oncogenic viruses are usually carried out on mouse models and in this framework viral antigens have shown to be relevant for tumour rejection. The most common oncoviruses employed in animal studies include papovaviruses, such as polyomavirus and simian virus 40 (SV40), and adenoviruses [54, 55, 59, 102, 130].

Oncofetal Antigens

Oncofetal antigens include the carcino-embryonic antigen (CEA) [44, 49], the breast cancer mucin MUC-1 [33, 86], the prostate-specific membrane antigen (PSMA) [72, 88, 131] and alpha fetoprotein (AFP) [20, 113]. These are proteins that are expressed at high levels on cancer cells and in normal developing (fetal) but not adult tissues. It is believed that the genes encoding the antigens are silenced during development and are derepressed upon malignant transformation. As techniques for detecting these proteins have improved, it has become clear that their expression in adults is not limited to tumours. The proteins are increased in tissues and in the circulation in various inflammatory conditions and are found in small quantities even in normal tissues. No evidence has shown that oncofetal antigens are important inducers of antitumour immunity [1].

Altered Glycolypid and Glycoprotein Antigens

Various aspects of the malignant phenotype associated with cancer cells are often related to an alteration of the biochemical properties of the cell membrane [10, 110, 123, 136]. Many researchers have reported a correlation of the processes of tumour invasion and metastasis with the expression of altered, immunogenic forms of membrane glycolipids and glycoproteins, including gangliosides, blood group antigens and mucins. Many of these antigens are tumour-specific and may be used as diagnostic markers and targets for therapy.

Tissue-Specific Differentiation Antigens

Tissue-specific differentiation antigens include PSA, MART-1/Melan-A, tyrosinase, gp100 and TRP-1 (gp75) [120]. These are proteins expressed in a tumour of a given type as well as in the normal tissue from which the tumour is derived [32, 43, 85, 99, 115, 118]. They are called differentiation antigens because they are specific for particular lineages or differentiation stages of various cell types. Their importance is as potential targets for immunotherapy and for identifying the tissue of origin of the tumours. According to [113] the realization of the existence of tissue-specific differentiation antigens came with the finding in [7] that some melanoma reactive CTLs recognized normal melanocytes. This suggested that tissue specific antigens might serve as the targets of tumour-reactive T-cells, a hypothesis that was confirmed by the isolation of a gene that was termed MART-1 or Melan-A following the screening of a cDNA library with HLA-A2 restricted tumour-infiltrating lymphocytes in [56] or CTL clones in [25] respectively.

2.5 Immunotherapy and Cancer Vaccines

According to [47] the idea of cancer vaccination can be traced back to the beginning of the twentieth century. At that time, the pre-existing clinical experience with regard to the basic principles of vaccination was mature enough to indicate that autologous or allogeneic tumour cells could be effective as therapeutic vaccines. The early considerations in this direction led to interesting observations such as the correlation of the effectiveness of the vaccine in use with a small tumour burden as well as with an increase in leukocyte counts reported in [138] as early as the year 1914.

Cancer vaccination differs from the usual practice of vaccination against infectious diseases mainly in two respects. First, as already indicated, cancer vaccination is of a therapeutic nature in contrast to the usual vaccination procedures, where the aim is the prevention of the development of a possible infection by the disease. Nonetheless, as noted in [47], some therapeutic vaccines for infectious diseases exist with the most representative of them being the vaccines against malaria [92] and leprosy [103]. Second, preventative vaccines work by boosting the immune system to expand memory cell clones against the antigens associated with an infectious disease – this will allow the immune system to respond rapidly in a possible future infection by the disease. In contrast the main objective of cancer vaccination is the activation of the effector components of the immune system of a tumour-bearing host so as to eradicate the tumour. Nevertheless, the establishment of memory is critical in this case as well in order for the host's immune system to be able to respond effectively in a possible future tumour recurrence.

Many different protocols have been developed for the administration of cancer vaccines in patients depending on various factors. For instance, when cancer vaccines are used in an adjuvant setting (i.e. treatment of patients after surgical removal of all clinically evident disease) the vaccinations are carried out over a period of 1–3 years with repeated immunizations in the first six months and less frequent injections in the remainder of years 1, 2 and 3 in order to avoid any possible side effects of the vaccine and to establish memory [47]. However, to date, *"there is no agreement as to the optimal frequency or of how long the vaccines should be administered"* [46]. In this direction, many authors have pointed out that mathematical and computational modelling can be of extreme importance in establishing a framework for the design and optimization of effective protocols [2, 3, 36, 63].

Although there is a lot of controversy concerning the administration of cancer vaccines, there are, nonetheless, some general principles which are guiding the development of modern protocols. For instance, it is experimentally observed that tumours have the ability to develop a variety of mechanisms for inhibiting or even evading the immune system responses [117]. Table 1 lists

Table 1. Mechanisms involved in inhibition of immune responses to tumours. Adapted from Hersey and Marincola (2002)

Mechanism	Factors Involved
Inhibition of antigen presentation	VEGF, IL-10
Inhibition of cytokine production	IL-10, TGF-β, α-MSH
Tolerance/anergy of T-cells	H_2O_2, TGF-β, Muc-1, α-MSH
Shift of TH_1TC_1 to TH_2TC_2	IL-10, TGF-β
Inhibition of migration of leukocytes from blood vessels	PGE_2, tumour matrix, P16E
Tumour-mediated destruction of T-cells	FasL, Muc-1
Resistance of tumour cells to killing	IL-10, immunoselection of MHC and antigen loss variants

some of the most common of these mechanisms along with the chemical factors involved. According to [47], "*specific inhibitors of some of these factors may be determined over the next few years but until more is known about the inhibiting factors, it appears appropriate to reduce the negative effects of tumours on immune responses by surgical removal of as much tumour as possible and immunization at sites removed from negative effects of the tumour*".

In the next section we will present our mathematical model for the immune response of cytotoxic T-lymphocytes to a solid tumour. The work is essentially a synthesis of that of [83, 84], with further, more detailed analysis of the model being carried out in [82].

3 Modelling the Spatio-temporal Response of Cytotoxic T-lymphocytes to a Solid Tumour

3.1 Deriving the Model

We consider a simplified process of a small, growing, avascular tumour that elicits a response from the host immune system and attracts a population of lymphocytes. The growing tumour is directly attacked by TICLs [52, 53, 57] which, in turn, secrete soluble diffusible factors (chemokines). These factors enable the TICLs to respond in a chemotactic manner (in addition to random motility) and migrate towards the tumour cells. Our model will therefore consist of six dependent variables denoted E, T, C, E^*, T^* and α, which are the local densities/concentrations of TICLs, tumour cells, TICL-tumour cell complexes, inactivated TICLs, "lethally hit" (or "programmed-for-lysis") tumour cells, and a single (generic) chemokine respectively.

We first of all consider the local interactions between the TICLs and tumour cells in vivo which may be described by the simplified kinetic scheme given in Fig. 3 (see also [65, 83]). The parameters k_1, k_{-1} and k_2 are non-negative kinetic constants: k_1 and k_{-1} describe the rate of binding of TICLs to tumour cells and detachment of TICLs from tumour cells *without* damaging cells; k_2 is the rate of detachment of TICLs from tumour cells, resulting in an irreversible programming of the tumour cells for lysis (i.e. death) with probability p or inactivating/killing TICLs with probability $(1 - p)$. For the first time, the possibility of a direct "counterattack" against the effector immune cells was theoretically postulated by Kuznetsov in his modeling of the local

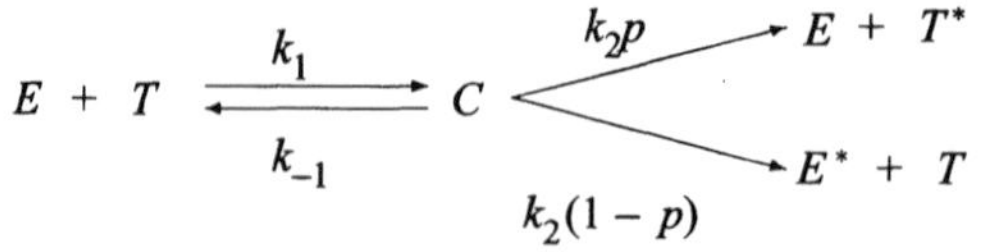

$$E + T \underset{k_{-1}}{\overset{k_1}{\rightleftharpoons}} C \overset{k_2 p}{\underset{k_2(1-p)}{\diagdown}} \begin{matrix} E + T^* \\ E^* + T \end{matrix}$$

Fig. 3. Schematic diagram of local lymphocyte-cancer cell interactions

interaction of cytotoxic lymphocytes and tumour cells in vivo [61]. Recently, it has been shown that such a mechanism might be realized through the Fas receptor (Fas, Apo-1/CD95) and its ligand (FasL, CD95L) [94]. Engagement of Fas on a target cell by FasL triggers a cascade of cellular events that result in programmed-cell-death. Both these transmembrane proteins (belonging to the tumour necrosis factor (TNF) family of receptors and ligands) are expressed on the surface of immune cells, including T-lymphocytes and NK-cells. However, many non-lymphoid tumour cells also express FasL which can counterattack and kill the Fas-sensitive tumour-infiltrating lymphocytes. On the other hand, most cancer cells, unlike normal cells, are relatively resistent to Fas-mediated apoptosis by the immune cells. Resistance to programmed-cell-death (apoptosis) through the Fas receptor pathway coupled with expression of the Fas ligand might enable many cancer cells to deliver a "counterattack" against attached cytotoxic lymphocytes.

Using the law of mass action, the above kinetic scheme can be "translated" into a system of ordinary differential equations

$$\frac{dE}{dt} = -k_1 ET + (k_{-1} + k_2 p)C \, , \tag{1}$$

$$\frac{dT}{dt} = -k_1 ET + (k_{-1} + k_2(1-p))C \, , \tag{2}$$

$$\frac{dC}{dt} = k_1 ET - (k_{-1} + k_2)C \, , \tag{3}$$

$$\frac{dE^*}{dt} = k_2(1-p)C \, , \tag{4}$$

$$\frac{dT^*}{dt} = k_2 pC \, . \tag{5}$$

Next we consider other kinetic interaction terms between the variables and examine migration mechanisms for the TICLs, tumour cells and also consider diffusion of the chemokines. We assume that there is no "nonlinear" migration of cells and no nonlinear diffusion of chemokine i.e. all random motility, chemotaxis and diffusion coefficients are assumed constant.

Tumour-Infiltrating Cytotoxic Lymphocytes

We assume that the TICLs have an element of random motility and also respond chemotactically to the chemokines. There is a source term modelling the underlying TICL production by the host immune system, a linear decay (death) term and an additional TICL proliferation term in response to the presence of the tumour cells. Combining these assumptions with the local kinetics (derived from Fig. 3) we have the following PDE for TICLs:

$$\frac{\partial E}{\partial t} = \overbrace{D_1 \nabla^2 E}^{\text{random motility}} \overbrace{-\chi \nabla \cdot (E \nabla \alpha)}^{\text{chemotaxis}} + \overbrace{s\,h(\mathbf{x})}^{\text{supply}} + \overbrace{\frac{fC}{g+T}}^{\text{proliferation}}$$

$$\overbrace{-\,d_1 E}^{\text{decay}} \overbrace{-\,k_1 ET + (k_{-1} + k_2 p)C}^{\text{local kinetics}},\tag{6}$$

where D_1, χ, s, f, g, d_1, k_1, k_{-1}, k_2, p are all positive constants. D_1 is the random motility coefficient of the TICLs and χ is the chemotaxis coefficient. The parameter s represents the "normal" rate of flow of mature lymphocytes into the tissue (non-enhanced by the presence of tumour cells). The function $h(\mathbf{x})$ is a Heaviside function, which aims to model the existence of a subregion of the domain of interest where initially there are only tumour cells and where lymphocytes do not reside. This region of the domain is penetrated by effector cells subsequently through the processes of diffusion and chemotaxis only (see below for a full discussion regarding this assumption).

The proliferation term $fC/(g+T)$, which has been introduced in [62], represents the experimentally observed enhanced proliferation of TICLs in response to the tumour and has been derived through data fitting (see also [65]). This functional form is consistent with a model in which one assumes that the enhanced proliferation of TICLs is due to signals, such as released interleukins, generated by effector cells in tumour cell-TICL complexes. We note that the growth factors that are secreted by lymphocytes in complexes (e.g IL-2) act mainly in an autocrine fashion. That is to say they act on the cell from which they have been secreted and thus, in our spatial setting, their action can be adequately described by a "local" kinetic term only, without the need to incorporate any additional information concerning diffusivity.

Chemokine Concentration

Chemokines are a super-family of small proteins (8-11kD) secreted primarily by leukocytes characterized by a few conserved cystein motifs. Expression of different cytokines and chemokines in tumour tissue (i.e. via gene delivery or the tumour tissue micro-environment) can induce host responses including in-filtration of T-cells capable of rejecting immuno-genic tumours [69]. However the production of chemokines in tumour tissue as well as the trafficking of lymphocytes into tumour tissue are dynamic, multi-step processes and currently the precise role of chemokines in tumour growth is still controversial [9, 119].

We assume that the chemokines are produced when lymphocytes are activated by tumour cell-TICL interactions. Thus we define chemokine production to be proportional to tumour cell-TICL complex density C. Once produced the chemokines are assumed to diffuse throughout the tissue and to decay in a simple manner with linear decay kinetics. Therefore the PDE for the chemokine concentration is:

$$\frac{\partial \alpha}{\partial t} = \overbrace{D_2 \nabla^2 \alpha}^{\text{diffusion}} + \overbrace{k_3 C}^{\text{production}} - \overbrace{d_4 \alpha}^{\text{decay}} , \tag{7}$$

where D_2, k_3, d_4 are positive parameters.

Tumour Cells

For a simplified description of the spatio-temporal growth of a solid tumour in the very early stages of its development, we will use a basic reaction-diffusion equation. On the kinetic level, the growth dynamics of solid tumours may be described adequately by the logistic equation:

$$\frac{dT}{dt} = b_1(1 - b_2 T)T , \tag{8}$$

which takes into account a density limitation of growth [65, 80, 81]. The inclusion of a spatial diffusion term in (8) leads to the well-known Fisher-Kolmogorov equation:

$$\frac{\partial T}{\partial t} = D_3 \nabla^2 T + b_1(1 - b_2 T)T , \tag{9}$$

which has been used by a number of authors for the modelling of the spatio-temporal evolution of solid tumours [30, 71]. In particular, the appropriateness of (9) for modelling tumour growth has been discussed in [30], where a lattice-free single-cell-based model of tumour growth in situ has been developed. Within realistic ranges of model parameters, the authors were able to provide a quantitative description of the growth curves in certain experiments. Furthermore, they have approximated the spatio-temporal evolution of their discrete model with a Fisher-Kolmogorov equation.

An alternative approach is to modify the logistic growth kinetics by incorporating terms modelling competition for space between various cell types [39, 40]. However, in the framework of our model, we will assume that the TICLs do not compete with the tumour cells for space. This is a reasonable assumption since according to observations [68] the volume of extracellular space in tumours is typically in the range 25–65% of the total volume of cells and hence there is enough space for the migration of lymphocytes within a tumour. Also, tumour cells lack the contact inhibition properties of normal cells and destroy the extracellular matrix. This allows the lymphocytes to migrate into the tumour tissue faster than in normal tissue, which has regular extracellular matrix. Therefore we do not explicitly include a term for space competition between the tumour cells and the lymphocytes and thus a logistic growth term is, we believe, a good first modelling approximation to the tumour growth local kinetics.

We assume that migration of the tumour cells may be described by simple random motility and hence the PDE governing the evolution of the tumour cell density is:

$$\frac{\partial T}{\partial t} = \overbrace{D_3 \nabla^2 T}^{\text{random motility}} + \overbrace{b_1(1 - b_2 T)T}^{\text{logistic growth}} \overbrace{-k_1 ET + (k_{-1} + k_2(1 - p))C}^{\text{local kinetics}},$$

(10)

where D_3 is the random motility coefficient of the tumour cells, and b_1, b_2, k_1, k_{-1}, k_2, p are positive parameters. The maximal growth rate of the tumour cell population is b_1, which incorporates both cell multiplication (mitosis) and death. The maximum density of the tumour cells is defined, and is represented by the parameter b_2^{-1} (cf. [31, 106]).

Tumour Cell-TICL Complexes

We assume that there is no diffusion of the complexes, only interactions governed by the local kinetics derived from Fig. 1. The absence of a diffusion term is justified by the fact that formation and dissociation of complexes occurs on a time scale of tens of minutes, whereas the random motility of the tumour cells, for example, occurs on a time scale of tens of hours. Thus, the cell-cell complexes do not have time to move. Therefore the equation for the complexes is given by

$$\frac{\partial C}{\partial t} = \overbrace{k_1 ET - (k_{-1} + k_2)C}^{\text{local kinetics}}.$$

(11)

Inactivated TICLs and Dead Tumour Cells

We assume that inactivated and "lethally hit" cells are quickly eliminated from the tissue (for example, by macrophages) and do not substantially influence the immune processes being analyzed (a slightly more complicated model might consider the re-introduction of the inactivated TICLs at some later stage). Inactivated cells also do not migrate and therefore we have:

$$\frac{\partial E^*}{\partial t} = \overbrace{k_2(1 - p)C}^{\text{local kinetics}} - \overbrace{d_2 E^*}^{\text{decay}},$$

(12)

$$\frac{\partial T^*}{\partial t} = \overbrace{k_2 p C}^{\text{local kinetics}} - \overbrace{d_3 T^*}^{\text{decay}}.$$

(13)

Therefore the complete system is:

$$\frac{\partial E}{\partial t} = D_1 \nabla^2 E - \chi \nabla \cdot (E \nabla \alpha) + s\, h(\mathbf{x}) + \frac{fC}{g+T} - d_1 E$$
$$- k_1 ET + (k_{-1} + k_2 p)C , \tag{14}$$

$$\frac{\partial \alpha}{\partial t} = D_2 \nabla^2 \alpha + k_3 C - d_4 \alpha , \tag{15}$$

$$\frac{\partial T}{\partial t} = D_3 \nabla^2 T + b_1(1 - b_2 T)T - k_1 ET + (k_{-1} + k_2(1-p))C , \tag{16}$$

$$\frac{\partial C}{\partial t} = k_1 ET - (k_{-1} + k_2)C , \tag{17}$$

$$\frac{\partial E^*}{\partial t} = k_2(1-p)C - d_2 E^* , \tag{18}$$

$$\frac{\partial T^*}{\partial t} = k_2 pC - d_3 T^*. \tag{19}$$

It is easy to see that (18) and (19) are only coupled to the full system through the complexes C and that neither E^* nor T^* have any effect on the variable C. Thus, for the remainder of this chapter, it is sufficient to analyse (14), (15), (16) and (17) which essentially dictate the behavior of the complete system.

For the sake of simplicity, in what follows in this section we will consider the case of one-dimensional tumour growth. Later on, in Sect. 3.7, we will present some simulations concerning the case of radially symmetric "three-dimensional" growth and in Sect. 3.8 we will focus on an explicit 2-dimensional finite-element-method simulation.

The Heaviside function $h(\mathbf{x})$ introduced in (14) models the existence of a subregion of the domain of interest where lymphocytes do not reside and which is penetrated by effector cells through the processes of diffusion and chemotaxis only. For instance, consider the specific case of a tumour that appears below the outer surface of a tissue (e.g. in the basal cell layer of the epidermis) and propagates into deeper levels of the tissue, i.e. invades the dermis (vertical tumour growth). This account could describe a nodular malignant melanoma which has no clinically or histologically evident radial growth phase. Of course, a short radial growth phase presumably does exist but dermal invasion is assumed to occur so rapidly that a preinvasive stage is not apparent [79]. Invasive growth is extremely insidious and dangerous, giving rise to metastases or secondary tumours [23, 24]. Considering the host's immune system response to the invasive tumour growth just described, we should note that intra-epidermal lymphocytes constitute only about 2% of skin-associated lymphocytes (the rest reside in the dermis). Intra-epidermal T cells may express a more restricted set of antigen receptors than do lymphocytes in most extracutaneous tissues. In mice (and some other species), many intra-epidermal lymphocytes are T cells that express an uncommon type of antigen receptor formed by γ and δ chains instead of the usual α and β chains of the antigen receptors of CD4$^+$ and CD8$^+$ T cells. This is also true of intra-epithelial lymphocytes in the intestine. Neither the specificity nor the function of this T cell subpopulation is clearly defined [1]. Thus, for the purposes of our modelling, we can assume that intra-epidermal lymphocytes are not relevant to the evolution of our system. Therefore, we separate the domain of interest to two subregions, an epidermis-like one and a dermis-like one, by introducing the Heaviside function.

We note here that the one-dimensional version of (14), (15), (16) and (17) does not entirely capture the evolution of a malignant melanoma of the skin, since the actual geometry is more intricate and complicated. However our purpose here is to investigate the dynamics of the model under discussion in a simple one-dimensional setting, which can give interesting insights. Nevertheless, our setting can be modified towards more realistic geometries.

We define the one-dimensional spatial domain to be the interval $[0, x_0]$, and we assume that there are two distinct regions in this interval – one region entirely occupied by tumour cells, the other entirely occupied by the immune cells. We propose that an initial interval of tumour localization is $[0, l]$, where $l = 0.2x_0$. Therefore the function $h(x)$ (cf. 14) is defined as follows:

$$h(x) = \begin{cases} 0, & \text{if } x - l \leq 0, \\ 1, & \text{if } x - l > 0. \end{cases}$$

3.2 Boundary and Initial Conditions

We now close the system by applying appropriate boundary and initial conditions. Zero-flux boundary conditions (BC) are imposed on the variables E, α, T, which in our system are equivalent to

$$\mathbf{n} \cdot \nabla E = \mathbf{n} \cdot \nabla \alpha = \mathbf{n} \cdot \nabla T = 0. \tag{20}$$

The initial conditions (IC) are given by:

$$E(x, 0) = \begin{cases} 0, & \text{if } 0 \leq x \leq l, \\ E_0(1 - \exp(-1000(x - l)^2)), & \text{if } l < x \leq x_0, \end{cases}$$

$$T(x, 0) = \begin{cases} T_0(1 - \exp(-1000(x - l)^2)), & \text{if } 0 \leq x \leq l, \\ 0, & \text{if } l < x \leq x_0, \end{cases} \tag{21}$$

$$C(x, 0) = \begin{cases} 0, & \text{if } x \notin [l - \epsilon, l + \epsilon], \\ C_0 \exp(-1000(x - l)^2), & \text{if } x \in [l - \epsilon, l + \epsilon], \end{cases}$$

$$\alpha(x, 0) = 0, \forall x \in [0, x_0],$$

where

$$E_0 = \frac{s}{d_1}, \quad T_0 = \frac{1}{b_2}, \quad C_0 = \min(E_0, T_0), \quad 0 < \epsilon \ll 1. \tag{22}$$

Figure 4 depicts qualitatively the ICs described in (21) (after the non-dimensionalization of the next section), which shows a front of tumour cells

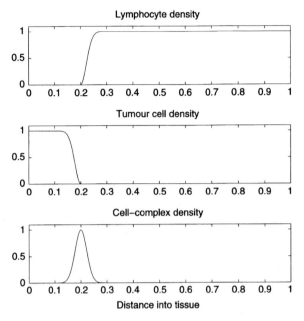

Fig. 4. Initial conditions used for the tumour infiltrating cytotoxic lymphocytes, tumour cells, and complexes

encountering a front of TICLs, resulting in the formation of TICL-tumour cell complexes. In the absence of a tumour, the homogeneous steady-state density of the TICLs is s/d_1 and therefore this is the value we have taken for the initial density E_0 of TICLs in the initial conditions. Similarly, in the absence of an immune response, the homogeneous steady-state density of the tumour cells is $1/b_2$ and this is what we take as the initial density of tumour cells T_0 in the initial conditions. Thus, when the fronts of the two cell populations meet, the maximum density of TICL-tumour cell complexes will be $\min(E_0, T_0)$ and hence our choice for C_0.

3.3 Estimation of Parameters

In order to carry out an analysis of the model by numerical methods it is useful to estimate values for the parameters obtained from experimental data and work with a non-dimensional system of equations.

The murine B cell lymphoma (BCL_1) is used as an experimental model of tumour dormancy in mouse [125, 134]. It has been demonstrated that $CD8^+$ T-cells are required for inducing and maintaining dormancy in BCL_1. In these experiments $CD8^+$ T cells are enhanced with anti-Id antibodies into inducing dormancy by secreting IFN-γ. A description of the growth kinetics of a BCL_1 lymphoma in the spleen of recipient mice, chimeric with respect to the Major Histocompatibility Complex (MHC) [125], was provided by the model of [65].

The kinetic parameters (obtained in [22]) were determined to have the following values:

$b_1 = 0.18$ day^{-1}, $b_2 = 2.0 \times 10^{-9}$ cells^{-1}cm ,
$k_1 = 1.3 \times 10^{-7}$ day^{-1}cells^{-1}cm, $k_{-1} = 24.0$ day^{-1} ,
$k_2 = 7.2$ day^{-1}, $p = 0.9997$,
$d_1 = 0.0412$ day^{-1}, $f = 0.2988 \times 10^8$ day^{-1}cells cm^{-1} ,
$g = 2.02 \times 10^7$ cells cm^{-1}, $s = 1.36 \times 10^4$ day^{-1}cells cm^{-1} ,

In addition to the kinetic parameters, we require estimates of the cell motility parameters. As we have seen in the previous sections, a tumour may be infiltrated by TICLs as a result of passive migration (random motility) or active transport (chemotaxis). In the first case, the random motility coefficient of the TICLs can be evaluated employing Einstein's formula:

$$D_1 = \frac{kT}{6\pi R_1 \eta} ,$$

where k is Boltzmann's constant, T is the temperature in degrees Kelvin, R_1 is the average radius of a TICL and η is the viscosity coefficient of the medium. With values of $T = 310$K ($37°$C), $R_1 = 4\,\mu$m and $\eta = \eta_{water}$, this gives an estimate of the TICL random motility coefficient $D_1 = 7.0 \times 10^{-5}$ cm^2 day^{-1}. This value is close to the random motility coefficient of CTLs in vitro obtained by [116], studying sequential killing of immobilised allogenic tumour cells by CTLs.

The random motility of tumour cells in tissue is conditioned largely by the replication of the cells and the growth of the tumour. The random motility coefficient may therefore be estimated from the following equation [106]:

$$D_3 = 4R_2^2 a ,$$

where R_2 is the average radius of a tumour cell and a is the rate of duplication of a tumour cell. Assuming $R_2 = 4 - 15\,\mu$m and $a = 0.1 - 1$ day^{-1}, we obtain a range of values $D_3 = 0.6 - 9 \times 10^{-6}$ cm^2 day^{-1}.

We know that TICLs are capable of infiltrating solid or lymphoma-like tumours rather rapidly [53, 75, 108, 126] and it is apparent, that if the movement of the TICLs and/or tumour cells is an active process induced by chemoattractants, the value of the diffusion coefficients of these cells may be appreciably greater and perhaps even reach approximately 10^{-2} cm^2 day^{-1} [38, 53]. Thus, the intervals of variation of the random motility coefficients may be large, depending upon the physical and biochemical properties of the surrounding tissue matrix and the concentration of various chemoattractants (chemokinesis). However in the simulations to be presented in the subsequent sections we assumed all random motility to be constant and took $D_1 = D_3 = 10^{-6}$cm^{2}day^{-1}.

Chemokines diffuse several orders of magnitude faster than cells. A reasonable range of values for the diffusion coefficient of the chemokine D_2 is:

$$10^{-4}\text{cm}^2\text{day}^{-1} \leq D_2 \leq 10^{-2}\text{cm}^2\text{day}^{-1}. \tag{23}$$

A more precise estimate for D_2 can be found from the data of [67], which are concerned with the motion of monoclonal antibodies (MCA). These results can be modified to account for the molecular mass of a typical chemokine (11 kD) and then can be combined with the Stokes-Einstein formula. This yields a value of $D_2 = 8 \times 10^{-3}\,\text{cm}^2\,\text{day}^{-1}$. However, in our simulations the above range of values for D_2, given by equation (23), were used.

The half-life of chemokines is around 60 days [109] and so we obtained an estimate for d_4 of $0.693/60 = 1.155 \times 10^{-2}\text{day}^{-1}$. We estimated the chemotactic response of the TICLs from data of macrophages in response to MCP-1 [96, 127]. From the range of estimates in these papers we chose a value of $1.728 \times 10^6\,\text{cm}^2\,\text{day}^{-1}\,\text{M}^{-1}$. The chemokine production parameter k_3 was estimated from data from several groups [21, 70, 87]. These data estimated the rate of production of chemokine (lymphotactin and IL-8) to be in the range of $20-3000$ molecules $\cdot$ cell^{-1}minute^{-1}.

Before proceeding with the numerical analysis, we non-dimensionalize our equations in the standard manner.

3.4 Non-Dimensionalization

We non-dimensionalise equations (14), (15), (16) and (17), the boundary conditions and initial conditions. An order-of-magnitude density scale is selected for the E, T and C cell densities, of E_0, T_0 and C_0, respectively, as suggested by the initial conditions. These are then given as $E_0 \approx 3.3 \times 10^5$ cells $\cdot$ cm^{-1}, $T_0 = 0.5 \times 10^9$ cells $\cdot$ cm^{-1} and $C_0 = E_0 \approx 3.3 \times 10^5$ cells $\cdot$ cm^{-1}.

The chemokine concentration is normalised through some reference concentration α_0 which we take to be 10^{-10}M [91]. Time is scaled relative to the diffusion rate of the TICLs, i.e. $t_0 = x_0^2\,D_1^{-1}$, and the space variable x is scaled relative to the length of the region under consideration (i.e. $x_0 = 1$ cm). Then, on making the following substitutions:

$$\bar{E} = \frac{E}{E_0}, \quad \bar{T} = \frac{T}{T_0}, \quad \bar{C} = \frac{C}{C_0}, \quad \bar{\alpha} = \frac{\alpha}{\alpha_0}, \quad \bar{x} = \frac{x}{x_0}, \quad \bar{t} = \frac{t}{t_0},$$

and omitting the bars for the sake of clarity, equations (14), (15), (16) and (17) may be re-written as:

$$\frac{\partial E}{\partial t} = \nabla^2 E - \gamma \nabla(E \nabla \alpha) + \sigma h(x) + \frac{\rho C}{\eta + T} - \sigma E - \mu E T + \epsilon C, \tag{24}$$

$$\frac{\partial \alpha}{\partial t} = \delta \nabla^2 \alpha + \kappa C - \xi \alpha, \tag{25}$$

$$\frac{\partial T}{\partial t} = \omega \nabla^2 T + \beta_1 (1 - \beta_2 T) T - \phi E T + \lambda C, \tag{26}$$

$$\frac{\partial C}{\partial t} = \mu E T - \psi C, \tag{27}$$

where

$$\sigma = \frac{st_0}{E_0} = d_1 t_0, \qquad\qquad \rho = \frac{ft_0 C_0}{E_0 T_0}, \qquad\qquad \mu = \frac{k_1 t_0 T_0 E_0}{C_0} = k_1 t_0 T_0 \ ,$$

$$\eta = \frac{g}{T_0}, \qquad\qquad \epsilon = \frac{t_0 C_0 (k_{-1} + k_2 p)}{E_0}, \qquad \omega = \frac{D_3 t_0}{x_0^2} = D_3 D_1^{-1},$$

$$\beta_1 = b_1 t_0, \qquad\qquad \beta_2 = b_2 T_0, \qquad\qquad \phi = k_1 t_0 E_0,$$

$$\lambda = \frac{t_0 C_0 (k_{-1} + k_2 (1 - p))}{T_0}, \quad \psi = t_0 (k_{-1} + k_2), \qquad \gamma = \frac{\chi \alpha_0 t_0}{x_0^2} = \chi \alpha_0 D_1^{-1},$$

$$\delta = \frac{D_2 t_0}{x_0^2} = D_2 D_1^{-1}, \qquad\qquad \kappa = \frac{k_3 t_0 C_0}{\alpha_0}, \qquad\qquad \xi = d_4 t_0.$$

After non-dimensionalization, the boundary conditions become:

$$\frac{\partial E}{\partial x}(0,t) = 0, \qquad \frac{\partial E}{\partial x}(1,t) = 0 \ ,$$

$$\frac{\partial \alpha}{\partial x}(0,t) = 0, \qquad \frac{\partial \alpha}{\partial x}(1,t) = 0 \ ,$$

$$\frac{\partial T}{\partial x}(0,t) = 0, \qquad \frac{\partial T}{\partial x}(1,t) = 0 \ ,$$

which then imply, assuming some smoothness of the solution and the form of equation (27),

$$\frac{\partial C}{\partial x}(0,t) = 0, \qquad \frac{\partial C}{\partial x}(1,t) = 0 \ .$$

Our initial conditions take the following form:

$$E(x,0) = \begin{cases} 0, & \text{if } 0 \le x \le l \ , \\ 1 - \exp(-1000(x - l)^2), & \text{if } l < x \le 1 \ , \end{cases}$$

$$T(x,0) = \begin{cases} 1 - \exp(-1000(x - l)^2), & \text{if } 0 \le x \le l \ , \\ 0, & \text{if } l < x \le 1 \ , \end{cases} \qquad (28)$$

$$C(x,0) = \begin{cases} 0, & \text{if } x \notin [l - \epsilon, l + \epsilon] \ , \\ \exp(-1000(x - l)^2), & \text{if } x \in [l - \epsilon, l + \epsilon] \ , \end{cases}$$

$$\alpha(x,0) = 0, \ \forall x \in [0, 1].$$

Values for all the non-dimensional parameters are obtained from the estimated dimensional parameters in paragraph 3.3. Concerning the chemokine production rate, we note that the data presented in paragraph 3.3 correspond to in vitro experimental settings and that we were unable to find any in vivo measurements in the literature. However, it seems reasonable to assume that not all complexes formed in vivo result in chemokine production and thus our choice for the value of parameter κ is slightly lower than the minimum value coming from the available in vitro data. Therefore, in the following simulations a value of $\kappa = 10^4$ has been chosen. This value of κ is of the same order of magnitude as that of ξ i.e. $\kappa \approx \xi$ and we note that this is also in line with the non-dimensional argument presented in [96]. All values of the non-dimensionalized parameters are given in [84] where a full description of the numerical scheme used can also be found.

By employing a numerical method, we obtain solutions for the above non-dimensionalized system, in the following section.

3.5 Numerical Simulation Results

The non-dimensionalized model was solved numerically using NAG routine D03PCF, which integrates systems of partial differential equations via the method of lines and a stiff ODE solver. We are aware of the numerical difficulties that the model poses and special care was taken to specify an appropriate number of grid points used in the numerical scheme. The FORTRAN code, which sets up the system and calls the D03PCF routine, can be found in [84]. As noted above, the non-dimensionalized parameter values employed in the numerical simulations to be presented here are also explicitly provided within the code.

Figures 5(a)–(d) show the spatial distribution of TICL density within the tissue at times corresponding to 100, 400, 700 and 1000 days respectively. The figures show a heterogeneous spatial distribution of TICL density throughout the tissue. Figures 6(a)–(d) show the corresponding spatial distribution of tumour cell density within the tissue at times corresponding to 100, 400, 700 and 1000 days. The figures show a train of solitary-like waves invading the tissue and subsequently creating a spatially heterogeneous distribution of tumour cell density throughout. Figures 7(a)–(d) show the corresponding spatial distribution of tumour cell-lymphocyte complexes within the tissue at times corresponding to 100, 400, 700 and 1000 days respectively. The dynamics of this spatio-temporal heterogeneity appear to persist as the long-time behaviour of the system in Figs. 8–10 shows. The times here correspond to 3000, 5000, 7000 and 10000 days respectively.

In addition to observing the above spatio-temporal distributions of each cell type within the tissue, the temporal dynamics of the overall populations of each cell type (i.e total cell number) was examined. This was achieved by calculating the total number of each cell type within the whole tissue space using numerical quadrature. Figure 11(a) shows the variation in the number of TICLs within the tissue over time (approximately 80 years, an estimated average lifespan). Initially, the total number of TICLs within the tissue increases and then subsequently oscillates around some stationary level (approximately 5.9×10^6 cells). Long-time numerical calculations indicated that this behaviour will persist for all time.

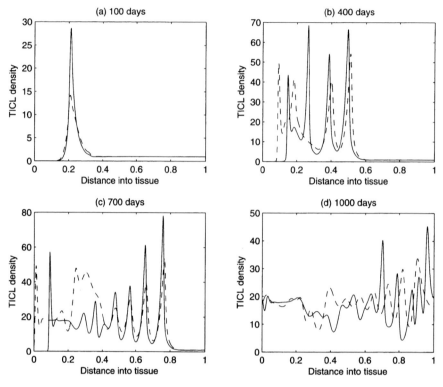

Fig. 5. Spatial distribution of TICL density within the tissue at times corresponding to 100, 400, 700 and 1000 days respectively. *Solid* line with chemotaxis, *dashed* without (i.e. $\gamma = 0$)

A similar scenario is observed for the tumour cell population. From Fig. 11(b), we observe that initially, the tumour cell population decreases in number before subsequently oscillating around some stationary value (approximately 10^7 cells) for all time. Figure 11(c) gives the corresponding temporal dynamics of the complexes. Figure 12 provides a more detailed view of the early oscillations in the total number of tumour cells.

The above simulations appear to indicate that eventually the tumour cells develop very small-amplitude oscillations about a "dormant" state, indicating that the TICLs have successfully managed to keep the tumour under control. The numerical simulations demonstrate the existence of cell distributions that are quasi-stationary in time and heterogeneous in space.

Concerning the spatial evolution of cancer dormancy with reference to the aspect of spatial containment, we note that the use of a fixed domain is consistent with various realistic biological settings. BCL_1 lymphomas of the spleen, for instance, are considered to be very good in vivo experimental models for investigating the various aspects of tumour development precisely

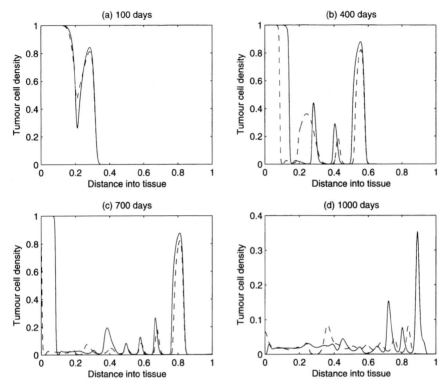

Fig. 6. Spatial distribution of tumour cell density within the tissue at times corresponding to 100, 400, 700 and 1000 days respectively. *Solid* line with chemotaxis, *dashed* without (i.e. $\gamma = 0$)

due to the fact that tumour cells are spatially contained within the lymph tissue of the spleen. Spleens in mice are elongated organs with boundaries defined by very strong basal membranes, which do not permit the tumour cells to escape unless they break these membranes (through well-known invasive processes) and then initiate metastases. However, in our model, we do not consider these cases and this is why we employ a fixed domain and impose zero-flux boundary conditions. Of course, if the domain itself were evolving the tumour cells would not be contained in space, but would rather spread throughout the domain. In the latter case we note that, from a mathematical point of view, it would be trivial to induce some kind of spatial containment of the tumour cells in a subregion of the domain by incorporating some non-autonomous ODE kinetics. However, this is not a realistic approach for the biological settings we consider and our numerical simulations do reflect several temporal as well as spatial aspects of tumour dormancy as these are described in various immunomorphological investigations.

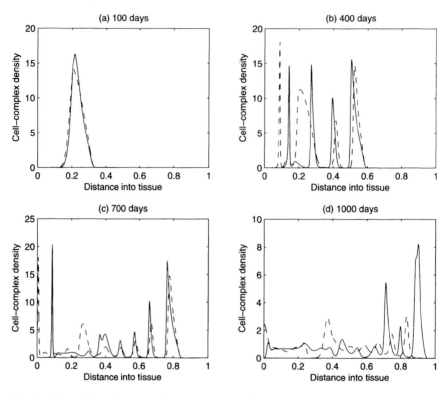

Fig. 7. Spatial distribution of tumour cell-TICL complex density within the tissue at times corresponding to 100, 400, 700 and 1000 days respectively. *Solid* line with chemotaxis, *dashed* without (i.e. $\gamma = 0$)

The interesting spatio-temporal dynamics of the system (i.e. the irregular invasive "waves") require us to investigate the underlying (spatially homogeneous) kinetics of our system. The interested reader is referred to the papers of [82–84] where a full analysis (bifurcation analysis and travelling-wave analysis) is presented.

3.6 Cancer Vaccination

In this section we consider the spatio-temporal dynamics of system (24)–(27) in the framework of a cancer-vaccination scenario. More precisely, we modify the Heaviside function so as to enable an increased influx of effector cells in a sub-region of the domain of interest. As in Sect. 3.5, we consider a one-dimensional domain, which, after non-dimensionalization, is identified by the interval $[0, 1] \subset \mathbb{R}$. The (non-dimensionalized) modified Heaviside function employed, hereafter called the influx funtion, is given by

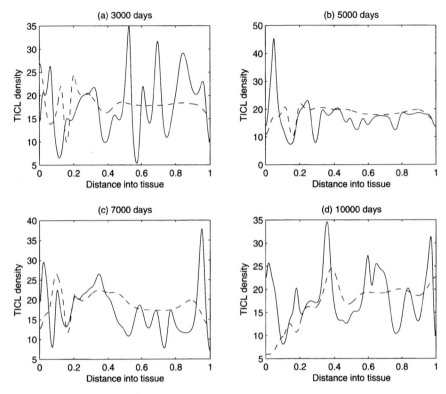

Fig. 8. Spatial distribution of TICL density within the tissue at times corresponding to 3000, 5000, 7000 and 10000 days respectively. *Solid* line with chemotaxis, *dashed* without (i.e. $\gamma = 0$)

$$h(x) = \begin{cases} 0, & \text{if } 0 \leq x \leq 0.2 , \\ 1, & \text{if } 0.2 < x < 0.8 , \\ 100, & \text{if } 0.8 \leq x \leq 1 . \end{cases}$$

Clearly, according to equation (24), this choice leads to an enhanced influx of effector cells into the interval [0.8, 1]. This could be due to the local administration of vaccine agents, which enhance the influx of effector cells into the region, or alternatively due to the direct administration of activated cytotoxic T-lymphocytes in an adoptive immunotherapy context [1, 47, 63].

The modified system of equations (24)–(27), with the influx function incorporated, was solved numerically over the interval [0, 1] with zero-flux boundary conditions imposed and the initial conditions given by (28). The results of the numerical simulations are depicted in Figs. 13–15, where we neglect the early transient dynamics by focusing on a time interval between 1000 and 1500 days. The spatio-temporal dynamics of the system over the interval [0, 0.8) are similar to the dynamics discussed in Sect. 3.5. However, as can be seen in Figs. 14 and 15, the enhanced influx of effector cells leads to the existence of

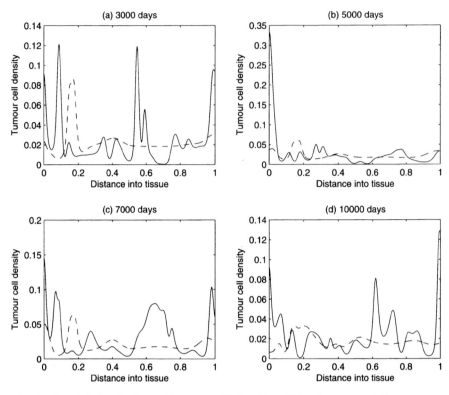

Fig. 9. Spatial distribution of tumour cell density within the tissue at times corresponding to 3000, 5000, 7000 and 10000 days respectively. *Solid* line with chemotaxis, *dashed* without (i.e. $\gamma = 0$)

a sub-region of the domain of interest where the tumour cells are effectively eradicated, resulting in the spatial containment of the tumour mass within the limits of a well-defined restricted domain. Moreover, as Fig. 13 shows, the tumour-free sub-region is characterized by an increased number of effector cells, with a stationary-in-time effector cell distribution. Long-time numerical computations indicated that this behaviour will persist for all time.

3.7 Radially Symmetric Solid Tumour Growth

We now turn our attention to the numerical solution of the system of equations (24), (26), and (27) in a radially symmetric 3-dimensional setting. In particular, we seek solutions of the form $E(r,t)$, $T(r,t)$, and $C(r,t)$ where r is the radius in spherical polar coordinates. In this setting we are assuming that the growth of the solid tumour represents the early avascular phase observed in multicell spheroids. We assume that there is no necrotic core only viable, proliferating cells. We also study the case where chemotaxis is not present

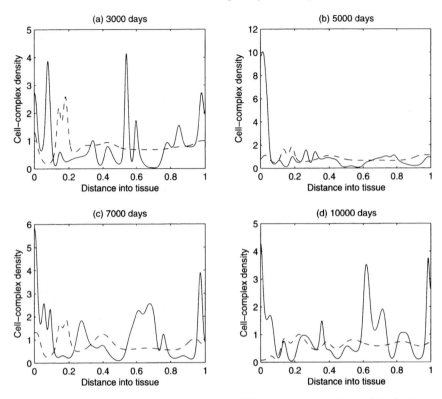

Fig. 10. Spatial distribution of tumour cell-TICL complex density within the tissue at times corresponding to 3000, 5000, 7000 and 10000 days respectively. *Solid* line with chemotaxis, *dashed* without (i.e. $\gamma = 0$)

(i.e. $\gamma = 0$). Rewriting the system in terms of spherical coordinates (assuming that all the partial derivatives of E and T with respect to the spherical polar angles θ and ϕ are equal to zero) we have:

$$\frac{\partial E}{\partial t} = \frac{1}{r^2}\left[\frac{\partial}{\partial r}\left(r^2\frac{\partial E}{\partial r}\right)\right] + \sigma h(r) + \frac{\rho C}{\eta + T} - \sigma E - \mu ET + \epsilon C\,, \quad (29)$$

$$\frac{\partial T}{\partial t} = \frac{1}{r^2}\left[\frac{\partial}{\partial r}\left(\omega r^2\frac{\partial T}{\partial r}\right)\right] + \beta_1(1 - \beta_2 T)T - \phi ET + \lambda C\,, \quad (30)$$

$$\frac{\partial C}{\partial t} = \mu ET - \psi C. \quad (31)$$

The results of the numerical simulations are presented in Figs. 16–18 which show cross sections through the spherical tumour. Figure 16 shows the spatial distribution of TICL density within the tissue at times corresponding to 100, 400, 700 and 1000 days respectively. The figures show a heterogeneous spatial distribution of TICL density throughout the tissue. Figure 17 shows the corresponding spatial distribution of tumour cell density within the tissue

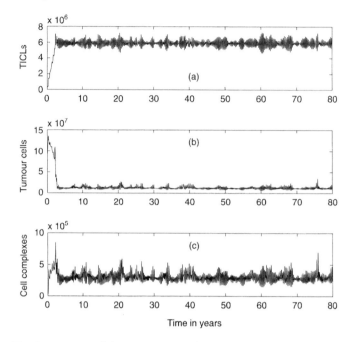

Fig. 11. Total number of (**a**) lymphocytes, (**b**) tumour cells, and (**c**) tumour cell-TICL complexes within tissue over a period of 80 years

at times corresponding to 100, 400, 700 and 1000 days. The figures show a train of solitary-like waves invading the tissue and subsequently creating a spatially heterogeneous distribution of tumour cell density throughout. Figure 18 shows the corresponding spatial distribution of tumour cell-lymphocyte complexes within the tissue at times corresponding to 100, 400, 700 and 1000 days respectively.

3.8 Explicit 2-Dimensional Modelling

In this section we undertake an explicit two-dimensional numerical investigation of a restricted version of the model developed in Sect. 3.1. More precisely, the full model is a system of a mixed hyperbolic-parabolic type and as such it poses various difficulties in its numerical approach, especially in a multi-dimensional framework. Hence, for the sake of computational simplicity, we do not consider the effect of chemotaxis and we omit the Heaviside function in (24). However, preliminary numerical experimentations with FEMLAB[3] on the restricted system over a rectangular two-dimensional domain indicated the necessity for either greatly enhancing the grid or using a stabilizing method, such as streamline diffusion for instance, since several difficulties associated

---

[3]See www.comsol.com

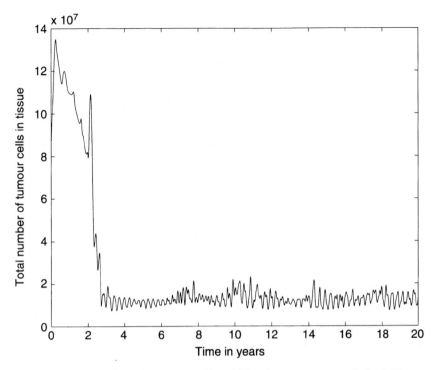

Fig. 12. Total number of tumour cells within tissue over a period of 20 years illustrating the early dynamics. The number of tumour cells initially decreases before settling down to an oscillatory behaviour around a stationary level of approximately 10^7 cells

with numerical instabilities emerged. We note here that this was anticipated since the model, even in its restricted form, combines two highly different time scales – the one associated with the slow spatial movement of the cells and the one that underlies the fast reaction kinetics. Nonetheless, the modelling assumptions, and in particular the estimated ranges of values for the random motility coefficients, suggested that we could also modify the system towards a more diffusive setting and thus we have chosen to alter the random motility coefficients in the non-dimensionalized system by multiplying them by a factor of ten. Specifically then, we have chosen to focus on the following non-dimensionalized reaction-diffusion system:

$$\frac{\partial E}{\partial t} = \omega_1 \nabla^2 E + \sigma + \frac{\rho C}{\eta + T} - \sigma E - \mu ET + \varepsilon C \ , \tag{32}$$

$$\frac{\partial T}{\partial t} = \omega_2 \nabla^2 T + \beta_1 (1 - \beta_2 T)T - \phi ET + \lambda C \ , \tag{33}$$

$$\frac{\partial C}{\partial t} = \mu ET - \psi C \ , \tag{34}$$

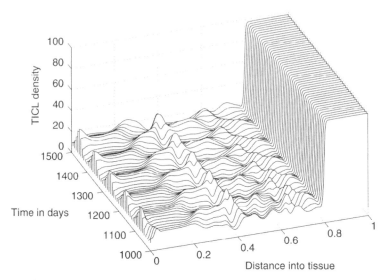

Fig. 13. Cancer-vaccination simulation. Spatial distribution of TICL density within the tissue at the time interval between 1000 and 1500 days

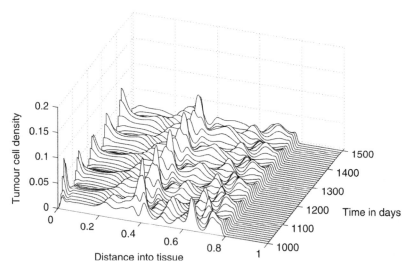

Fig. 14. Cancer-vaccination simulation. Spatial distribution of tumour cell density within the tissue (between 1000 and 1500 days)

where σ, ρ, η, μ, ε, β_1, β_2, ϕ, λ and ψ are the non-dimensionalized parameters defined previously, and $\omega_1 = \omega_2 = 10$.

In what follows, we consider the system (32)–(34) over the two-dimensional rectangular domain

$$D = [0, 1] \times [0, 1] \subset \mathbb{R}^2,$$

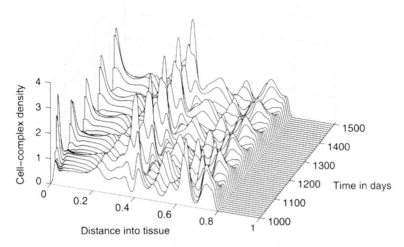

Fig. 15. Cancer-vaccination simulation. Spatial distribution of tumour cell-TICL complex density within the tissue (between 1000 and 1500 days)

with zero-flux boundary conditions imposed. Moreover, we assume that initially there are no complexes formed and that there is a homogeneous distribution of effector cells. That is to say, for all $(x, y) \in D$,

$$E(x, y, 0) = 1 \text{ and } C(x, y, 0) = 0 .$$

We also assume that three small clumps of tumour cells exist as Fig. 19 shows[4].

We have used FEMLAB to solve equations (32)–(34) with the boundary and initial conditions described. FEMLAB provides a number of options to the user concerning the discretization of partial differential equations and the approximation of solutions. We have specified to the software a finite element space based on triangular linear Lagrange elements. The mesh consisted of 36,090 nodes and 71,606 elements, 572 of which were associated with the boundary of the domain. FEMLAB controls the quality of the mesh by assigning to each element a quality measure – a number q between 0 and 1. In 2-D, the triangle quality measure employed by FEMLAB is given by:

$$q = \frac{4\sqrt{3}\alpha}{h_1^2 + h_2^2 + h_3^2} ,$$

where α is the area and h_1, h_2 and h_3 the side lengths of the triangle. Some information concerning the quality of the mesh we have used is provided in Fig. 20, which shows a histogram of the distribution of the measure q for all the elements in the mesh.

---

[4] An explicit algebraic expression for the initial tumour cell distribution can be found in the FEMLAB M-file in [84].

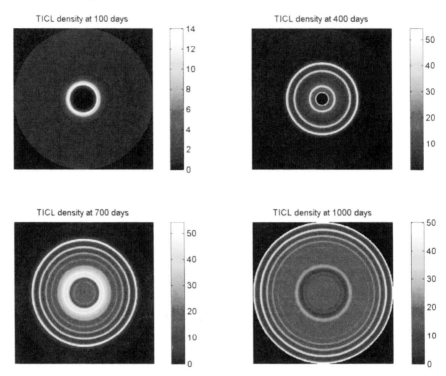

Fig. 16. Spatial distribution of TICL density within the tissue at times corresponding to 100, 400, 700 and 1000 days respectively

The above spatial finite element discretization was combined with a finite difference solver in a method-of-lines approach. In particular, we have used MATLAB's `ode15s` (see [122]) for the time integration of the associated semi-discrete Galerkin formulation of equations (32)–(34). The fast time scale underlying the kinetics of the model suggested that special care should be taken with respect to the time stepping and thus we have specified to the software a maximum (non-dimensional) time step of 10^{-7}.

The results of the numerical computations are shown in Figs. 21–26. In particular, Figs. 21(a)–(d) show the evolution of the (non-dimensionalized) spatial distribution of TICL density within the tissue at times corresponding to 80, 100, 200 and 400 days respectively. We note that these values correspond to 80×10^{-6}, 100×10^{-6}, 200×10^{-6} and 400×10^{-6} non-dimensional time units respectively. The corresponding spatial distributions of tumour cell and tumour cell-TICL complex densities are shown in Figs. 22(a)–(d) and 23(a)–(d). Similarly, Figs. 24(a)–(d) show time instances of the evolution of the spatial distribution of TICL density within the tissue at times corresponding to 480, 500, 640 and 700 days respectively, whereas Figs. 25(a)–(d) and 26(a)–(d) show the corresponding spatial distributions of tumour cell and tumour

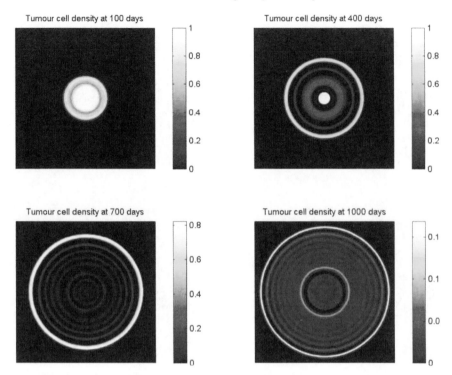

Fig. 17. Spatial distribution of tumour cell density within the tissue at times corresponding to 100, 400, 700 and 1000 days respectively

cell-TICL complex densities. Clearly, the simulations depict a reduction of the tumour bulk as a result of the cytotoxic activity of the TICLs, accompanied with an irregular evolution of the spatial distribution of the tumour cell density.

4 Discussion and Conclusions

In this chapter we have examined a spatio-temporal mathematical model describing the growth of a solid tumour in the presence of an immune system response. In particular, we focussed attention upon the interactions of tumour cells with a special sub-population of T-cells, so-called tumour-infiltrating cytotoxic lymphocytes (TICLs), in a relatively small, multicellular tumour, without central necrosis and at some stage prior to tumour-induced angiogenesis. The T-lymphocytes were assumed to migrate into the growing solid tumour and interact with the tumour cells in such a way that lymphocyte-tumour cell complexes were formed. These complexes resulted in either the death of the tumour cells (the normal situation) or the inactivation (sometimes even the death) of the lymphocytes. The migration of the TICLs was determined by a

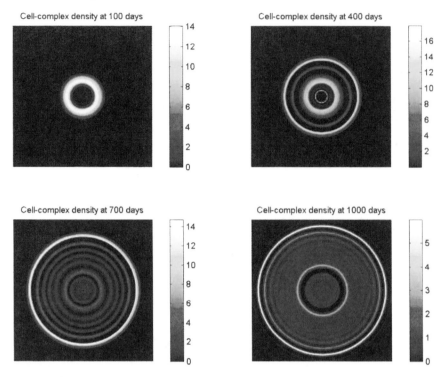

Fig. 18. Spatial distribution of tumour cell-TICL complex density within the tissue at times corresponding to 100, 400, 700 and 1000 days respectively

combination of random motility and chemotaxis in response to the presence of specialized chemoattractants (chemokines). The resulting system of four nonlinear partial differential equations (TICLs, tumour cells, complexes and chemokines) was analysed and numerical simulations were presented.

For a particular choice of parameters the model was able to simulate the phenomenon of cancer dormancy – a clinical condition that has been observed in breast cancers, neuroblastomas, melanomas, osteogenic sarcomas, and in several types of lymphomas – by depicting spatially unstable and heterogeneous tumour cell distributions that were nonetheless characterized by a relatively small total number of tumour cells. This behaviour was consistent with several immunomorphological investigations. However, as noted in [82], the alteration of certain parameters of the model is enough to induce bifurcations into the system, which in turn result in the existence of travelling-wave-like solutions in the numerical simulations. These travelling waves are of great importance because when they exist, the tumour invades the healthy tissue at its full potential escaping the host's immune surveillance. The existence of these travelling waves has been established rigorously in [82] for a reduced system, which nonetheless captures the essential elements of the full model.

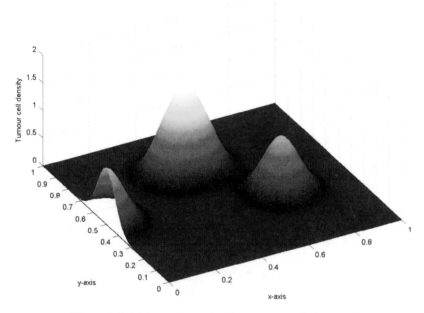

Fig. 19. Initial condition for the tumour cell density T

Various new results have been presented in this chapter and, in particular, the model was tested under various different biological settings (radially symmetric growth, 2-dimensional growth in a square domain, cancer vaccination etc.). The simulations presented here agree with the theoretical indications provided in [83] according to which the irregular evolution of the cancer dormancy simulations is an actual manifestation of spatio-temporal chaos. In this direction, a more in-depth bifurcation analysis of the ODE kinetics of the model reveals the existence of oscillatory solutions for the ODE system emerging through a Hopf bifurcation and establishes a correlation between the existence of the associated stable limit cycle with the irregular spatio-temporal evolution of the PDE system and the onset of cancer dormancy (see [82, 83]).

Our numerical and bifurcation analysis of the spatio-temporal model of cytotoxic T-cell dynamics in cancer tissue supports the idea that the TICLs can play an important role in the control of cancer dormancy. Moreover, the model allowed us to identify certain critical parameters of the process in which cancer cells are present in a tissue but do not clinically occur for a long period of time, but can begin to grow progressively at later date. Hence, our model could be potentially used to estimate the time interval between the primary treatment of an immunogenic tumour and tumour recurrence (see also [63]).

We note that heterogeneous spatial patterning in an immune-system model has also been found by [96–98] concerning macrophage interactions with tumour/mutant cells. In this case however the patterning was produced via an

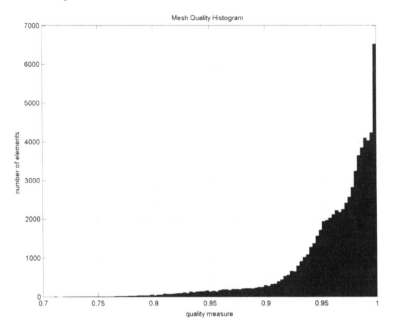

Fig. 20. Histogram showing the distribution of the mesh qualities for all elements in the mesh

activator-inhibitor (Turing) mechanism by considering the mutant cells as the local activator and the chemical regulator as the long-range inhibitor. There are some significant differences between that work and the one presented here. The kinetics of the model presented in [98] did not exhibit any Hopf bifurcation and, in the absence of a macrophage-based immunotherapy, the introduction of a small mutant cell density always caused the system to evolve to one of the two possible tumour invasion steady states it was able to predict. In our case, depending on the choice of parameters, the system kinetics may evolve to a tumour invasion steady state but they can also display an oscillatory behavior with the tumour cell density bounded as a result of the cytotoxic activity of the TICLs.

Perhaps more appropriately, the evolution of the formal kinetics of our system appears to have some similarities with the evolution of the ODE kinetics of the ecological models presented in [124]. In both cases the global dynamics concerning the positive solutions consist of two unstable steady states and a stable limit cycle emerging through a Hopf bifurcation (see [83]). However some differences between the models exist with the most obvious of them being the different biological frameworks (ecology vs. immunology) and the difference in the dimensions of the corresponding phase spaces. We note that previous authors have pointed out the existence of similarities between the immune system response to immunogenic antigens and predator-prey ecological interactions [93] – and more generally an "ecological competition" between

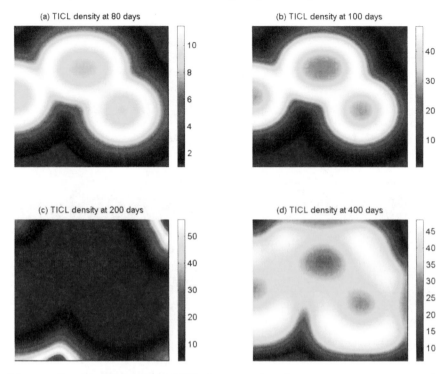

Fig. 21. Spatial distribution of TICL density within the tissue at times corresponding to 80, 100, 200 and 400 days respectively

cancer cells and normal tissue cells [39, 40] – and that the mathematical particularities of our system could lead to entirely different spatio-temporal dynamics than those presented in [124]. This is to be investigated.

The numerical predictions of our model make it possible to comprehend the mechanisms involved in the appearance of spatio-temporal heterogeneities detected in solid tumours infiltrated by cytotoxic lymphocytes. These are described in numerous immunomorphological investigations [14, 90]. We note that this model could be extended further. Specifically, explicit interactions between the cancer cells and the host tissue could be incorporated into the basic kinetic model (Fig. 3). For example, [139] proposed a mathematical model of a tumour cell "counter-attack" against cytotoxic T-cells. The model consists of ordinary differential equations which represent the cross-linking of FasL and Fas. The authors consider the antagonistic interactions of two cell types: armed effector T-cells, and FasL positive tumour cells. The model is based upon the observation that certain types of human tumours can produce functional FasL and can induce the apoptotic killing of activated lymphocytes in vitro.

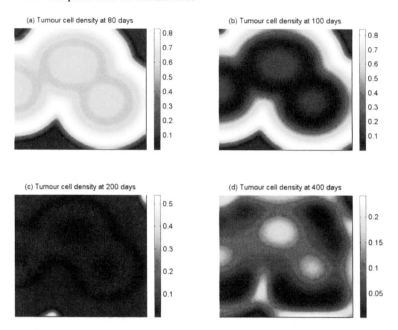

Fig. 22. Spatial distribution of tumour cell density within the tissue at times corresponding to 80, 100, 200 and 400 days respectively

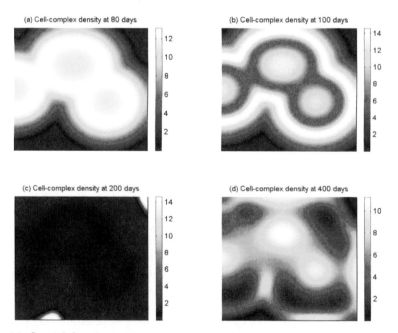

Fig. 23. Spatial distribution of tumour cell-TICL complex density within the tissue at times corresponding to 80, 100, 200 and 400 days respectively

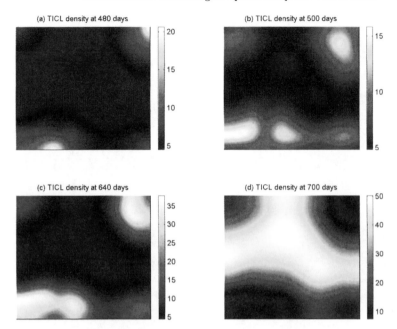

Fig. 24. Spatial distribution of TICL density within the tissue at times corresponding to 480, 500, 640 and 700 days respectively

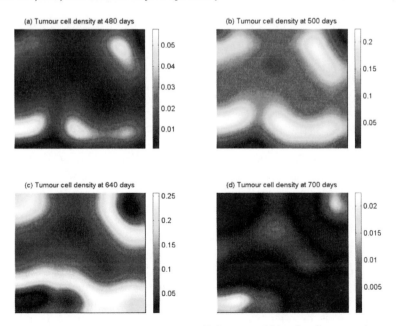

Fig. 25. Spatial distribution of tumour cell density within the tissue at times corresponding to 480, 500, 640 and 700 days respectively

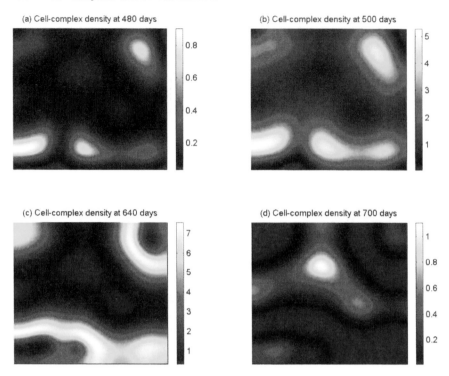

Fig. 26. Spatial distribution of tumour cell-TICL complex density within the tissue at times corresponding to 480, 500, 640 and 700 days respectively

Recent model-fitting predicts that the life time of effector T-cells in vivo could be short (about several days) [63]. Long term maintainance of anti-cancer immunity after stopping immunotherapy could be improved if long-life immune memory cells could be activated during immunization. In particular, numerical modelling by [63] suggests that immune memory T-killer cells could be critical targets for immunization and vaccination strategies against solid tumours. Thus, an incorporation of the memory cells in our model could be helpful in better understanding cancer dormancy and cancer re-growth mechanisms and in optimizing the therapeutic strategy to reduce the risk of tumour relapse.

Finally, the familiar concept of a central necrotic core (and explicit oxygen distribution/uptake) could also be incorporated. It has been stated that the rate of macrophage and neutrophil accumulation in a spheroid depends on the density of tumour cells and is determined by a law analogous to that of Michaelis-Menten kinetics, while the accumulation of immune lymphocytes in a tumour is determined by the three-cell cooperation of lymphocytes, macrophages and tumour cells (see [66]). This data could provide further adaptations to our model, incorporating new cell types and increasing the realism of the system.

We hope that the results presented here (and the effects caused by the nonlinearity of the system) will make it possible for researchers and clinicians to have a better idea of the complicated and sometimes counter-intuitive outcome of processes occurring in immune-system interactions with tumour cells and thereby to develop more effective immunotherapy strategies and treatments for the control and possible elimination of cancers. Our modelling and analysis offers the potential for quantitative analysis of mechanisms of tumour-cell-host-cell interactions and for the optimization of tumour immunotherapy and genetically engineered anti-tumour vaccines.

Acknowledgement

This work was supported by a Framework VI EU Human Resources and Mobility Marie Curie Research Training Network, MRTN-CT-2004-503661, "Modelling, Mathematical Methods and Computer Simulation of Tumour Growth and Therapy".

References

1. Abbas, A., Lichtman, A. & Pober, J. (2000), *Cellular and Molecular Immunology*, 4th edn, W.B. Saunders Company.
2. Adam, J. (1993), "The dynamics of growth-factor-modified immune response to cancer growth: One dimensional models", *Math. Comp. Modell.* **17**, 83–106.
3. Adam, J. & Bellomo, N., eds (1997), *A Survey of Models for Tumor-Immune System Dynamics*, Birkhaüser, Boston.
4. Alsabti, A. (1978), "Tumour dormant state", *Tumour Res.* **13**(1), 1–13.
5. Ambrosi, D., Bellomo, N. & Preziosi, L. (2002), "Modelling tumor progression, heterogeneity, and immune competition", *J. Theor. Medicine* **4**, 51–65.
6. Angelis, E. D., Delitala, M., Marasco, A. & Romano, A. (2003), "Bifurcation analysis for a mean field modelling of tumor and immune system competition", *Math. Comp. Modelling* **37**, 1131–1142.
7. Annichini, A., Maccalli, C., Mortarini, R., Salvi, S., Mazzocchi, A., Squarcina, P., Herlyn, M. & Parmiani, G. (1993), "Melanoma cells and normal melanocytes share antigens recognized by HLA-A2 restricted cytotoxic T cell clones from melanoma patients", *J. Exp. Med.* **177**, 989–998.
8. Baldwin, R. (1955), "Immunity to methylcholanthrene-induced tumors in inbred rats following atrophy and regression of implanted tumors", *Br. J. Cancer* **9**, 652–656.
9. Bar-Eli, M. (1999), "Role of interleukin-8 in tumor growth and metastasis of human melanoma", *Pathobiology* **67**(1), 12–18.
10. Barz, D., Goppelt, M., Schrimacher, V. & Resch, K. (1985), "Characterization of cellular and extracellular plasma membrane vesicles from a non-metastasizing lymphoma (Eb) and its metastasizing variant (Esb)", *Bioch. biophys. Acta* **814**, 77–84.

11. Bellomo, N. & Preziosi, L. (2000), "Modelling and mathematical problems related to tumor evolution and its interaction with the immune system", *Math. Comp. Modelling* **32**, 413–452.

12. Bellomo, N., Bellouquid, A. & Angelis, E. D. (2003), "The modelling of the immune competition by generalized kinetic (Boltzmann) models: Review and research perspectives", *Math. Comp. Modelling* **37**, 65–86.

13. Bellomo, N., Firmani, B. & Guerri, L. (1999), "Bifurcation analysis for a non-linear system of integro-difierential equations modelling tumor-immune cells competition", *Appl. Math. Letters* **12**, 39–44.

14. Berezhnaya, N. M., Yakimovich, L. V., Kobzar, R. A. S., Lyulkin, V. D. & Papivets, A. Y. (1986), "The effect of interleukin-2 on proliferation of explants of malignant soft-tissue tumours in diffusion chambers", *Experimental Oncology* **8**, 39–42.

15. Bohman, Y. (1976), *Metastases of corpus uterus cancer*, Leningrad: Medicine.

16. Breslow, N., Chan, C., Dhom, G., Drury, R., Franks, L., Gellei, B., Lee, Y., Lundberg, S., Sparke, B., Sternby, N. & Tulinius, H. (1977), "Latent carcinoma of prostate at autopsy in seven areas", *Intern. J. Cancer* **20**(5), 680–688.

17. Brocker, E., Zwaldo, G., Holzmann, B., Macher, E. & Sorg, C. (1988), "In-ammatory cell infiltrates in human melanoma at different stages of tumour progression", *Int. J. Cancer* **41**, 562–567.

18. Brunner, K., Mauel, J., Cerottini, J. & Chapuis, B. (1968), "Quantitative assay of the lytic action of immune lymphoid cells on 51-Cr-labelled allogeneic target cells in vitro, inhibition by isoantibody and by drugs", *Immunology* **14**, 181–196.

19. Burnet, F. M. (1970), "The concept of immunological surveillance", *Progress in Experimental Tumor Research* **13**, 1–27.

20. Butterfield, L., Koh, A., Meng, W., Vollmer, C., Ribas, A., Dissette, V., Lee, E., Glaspy, J., McBride, W. & Economou, J. (1999), "Generation of human T-cell responses to an HLA-A2.1-restricted peptide epitope derived from α-fetoprotein", *Cancer Res.* **59**, 3134–3142.

21. Cairns, C., Gordon, J., Li, F., Baca-Estrada, M., Moyana, T. & Xiang, J. (2001), "Lymphotaktin expression by engineered myeloma cells drives tumor regression: mediation by CD4+ and CD8+ T cells and neutrophils expressing XCR1 receptor", *J Immunol.* **167**(1), 57–65.

22. Chaplain, M., Kuznetsov, V., James, Z. & Stepanova, L. (1998), Spatiotemporal dynamics of the immune system response to cancer, *in* M. A. Horn, G. Simonett & G. Webb, eds, "Mathematical Models in Medical and Health Sciences", Vanderbilt University Press, pp. 1–20.

23. Clark, W. (1991), "Tumour progression and the nature of cancer", *Brit. J. Cancer* **64**, 631–644.

24. Clark, W., Elder, D. & Vanhorn, M. (1986), "The biologic forms of malignant melanoma", *Human Pathology* **17**, 443–450.

25. Coulie, P., Brichard, V., Pel, A. V., Wölfel, T., Schneider, J., Traversari, C., Mattei, S., Plaen, E. D., Lurquin, C., Szikora, J. et al. (1994), "A new gene coding for a differentiation antigen recognized by autologous cytolytic T lymphocytes on HLA-A2 melanomas", *J. Exp. Med.* **180**, 35–42.

26. Cowan, D., Hicks, K. & Wilson, W. (1996), "Multicellular membranes as an in vitro model for extravascular diffusion in tumours", *Br J Cancer Suppl.* **27**, S28–S31.

27. den Eynde, B. V. & van der Bruggen, P. (1997), "T cell defined tumor antigens", *Current Opinion in Immunology* **9**, 684–693.

28. Deweger, R., Wilbrink, B., Moberts, R., Mans, D., Oskam, R. & den Otten, W. (1987), "Immune reactivity in SL2 lymphoma-bearing mice compared with SL2-immunized mice", *Cancer Immun. Immunotherapy* **24**, 1191–1192.

29. Doherty, P., dunlop, M., Parish, C. & Zinkernagel, R. (1976), "Inammatory process in murine lymphocytic choriomeningitis is maximal in H-2K or H-2D compatible interactions", *J. Immunol.* **117**, 187–190.

30. Drasdo, D. & Höhme, S. (2003), "Individual-based approaches to birth and death in avascular tumours", *Math. Comput. Model.* **37**, 1163–1175.

31. Durand, R. & Sutherland, R. (1984), "Growth and cellular characteristics of multicell spheroids", *Recent Results in Cancer Research* **95**, 24–49.

32. Fernendez, N., Duffour, M., Perricaudet, M. et al. (1998), "Active specific T-cell based immunotherapy for cancers: nucleic acids, peptides, whole native proteins, recombinant viruses, with dendritic cell adjuvants or whole tumor cell-based vaccines. Principles and future prospects", *Cytokines Cell. Mol. Ther.* **4**(1), 53–65.

33. Finn, O., Jerome, K., Henderson, R. et al. (1995), "MUC-1 epithelial tumor mucin-based immunity and cancer vaccines", *Immunol. Rev.* **145**, 61–89.

34. Fisk, B., Blevins, T., Wharton, J. & Ioannides, C. (1995), "Identification of an immunodominant peptide of HER-2/neu protooncogene recognized by ovarian tumor-specific cytotoxic T lymphocytes lines", *J. Exp. Med.* **181**, 2109–2117.

35. Folkman, J. (1985), "How is blood-vessel growth regulated in normal and neoplastic tissue", *Proc. Amer. Assoc. Cancer Res.* **26**, 384–385.

36. Forni, G. (1996), "Tumor-host relationship: the viewpoint of an immunologist towards applied mathematicians", *Mathl. Comput. Modelling* **23**(6), 89–94.

37. Forni, G., Parmiani, G., Guarini, A. & Foa, R. (1994), "Gene transfer in tumour therapy", *Annals Oncol.* **5**, 789–794.

38. Friedl, P., Noble, P. & Zanker, K. (1995), "T-Lymphocyte locomotion in a 3-dimensional collagen matrix – expression and function of cell-adhesion molecules", *J. Immunol.* **154**, 4973–4985.

39. Gatenby, R. (1995), "Models of tumor-host intearcation as competing populations: Implications for tumor biology and treatment", *J. theor. Biol.* **176**, 447–455.

40. Gatenby, R. (1996), "Application of competition theory to tumour growth: Implications for tumour biology and treatment", *European Journal of Cancer* **32A**, 722–726.

41. Ghiso, J. A. (2002), "Inhibition of FAK signaling activated by urokinase receptor induces dormancy in human carcinoma cells in vivo", *Oncogene* **21**(16), 2513–2524.

42. Giovarelli, M., Cappello, P., Forni, G., Salcedo, T., Moore, P., LeFleur, D., Nardelli, B., Carlo, E., Lollini, P., Ruben, S., Ullrich, S., Garotta, G. & Musiani, P. (2000), "Tumor rejection and immune memory elicited by locally released LEC chemokine are associated with an impressive recruitment of APCs, lymphocytes and granulocytes", *J Immunol.* **164**(6), 3200–3206.

43. Greten, T. & Jaffee, E. (1999), "Cancer vaccines", *J. Clin. Oncol.* **17**(3), 1047–1060.

44. Hammarstrom, S. (1999), "The carcinoembryonic antigen (CEA) family: structures, suggested functions and expression in normal and malignant tissues", *Semin. Cancer Biol.* **9**, 67–81.

45. Hellström, K. & Hellström, I. (1969), "Cellular immunity against tumor antigens", *Adv. Cancer Res.* **12**, 167–223.
46. Hersey, P. (1997), "Melanoma vaccines. Prospects for the treatment of melanoma", *Expert Op., Invest. Drugs* **6**, 267–277.
47. Hersey, P. & Marincola, F. (2002), Immunotherapy of Cancer, *in* G. Parmiani & M. Lotze, eds, "Tumor Immunology: Molecularly Defined Antigens and Clinical Applications", Taylor & Francis, pp. 117–175.
48. Hesketh, R. (1997), *The Oncogene and Tumor Suppressor Genes Facts Book*, 2nd edn, Academic Press.
49. Hodge, J. (1996), "Carcinoembryonic antigen as a target for cancer vaccines", *Cancer Immunol. Immunother.* **43**, 127–134.
50. Holmberg, L. & Baum, M. (1996), "Work on your theories!", *Nat. Med.* **2**(8), 844–846.
51. Ikeda, H., Lethé, B., Lehmann, F., Baren, N. V., Baurain, J., Smet, C. D., Chambost, H., Vitale, M., Moretta, A., Boon, T. & Coulie, P. (1997), "Characterization of an antigen that is recognized on a melanoma showing partial HLA loss by CTL expressing an NK inhibitory receptor", *Immunity* **6**, 199–208.
52. Ioannides, C. & Whiteside, T. (1993), "T-cell recognition of human tumours – implications for molecular immunotherapy of cancer", *Clin. Immunol. Immunopath.* **66**, 91–106.
53. Jaaskelainen, J., Maenpaa, A., Patarroyo, M., Gahmberg, C., Somersalo, K., Tarkkanen, J., Kallio, M. & Timonen, T. (1992), "Migration of recombinant IL-2-activated T-cells and natural killer cells in the intercellular space of human H-2 glioma spheroids in vitro – a study on adhesion molecules involved", *J. Immunol.* **149**, 260–268.
54. Kast, W. & Melief, C. (1991), "Fine peptide specificity of cytotoxic T lymphocytes directed against adenovirus-induced tumors and peptide-MHC binding", *Int. J. Cancer Suppl.* **6**, 90–94.
55. Kast, W., Offringa, R., Peters, P., Voordouw, A., Meloen, R., van der Eb, A. & Melief, C. (1989), "Eradication of adenovirus E1-induced tumors by E1A specific cytotoxic T lymphocytes", *Cell* **59**, 603–614.
56. Kawakami, Y., Eliyahu, S., Delgado, C., Robbins, P., Rivoltini, L., Topalian, S., Miki, T. & Rosenberg, S. (1994), "Cloning of the gene coding for a shared human melanoma antigen recognized by autologous T cells infiltrated into tumor", *Proc. Natl. Acad. Sci. USA* **91**, 3515–3519.
57. Kawakami, Y., Nishimura, M., Restifo, N., Topalian, S., O"Neil, B., Shilyansky, J., Yannelli, J. & Rosenberg, S. (1993), "T-cell recognition of humanmelanoma antigens", *J. Immunotherapy* **14**, 88–93.
58. Kelly, C., Leek, R., Byrne, H., Cox, S., Harris, A. & Lewis, C. (2002), "Modelling macrophage infiltration into avascular tumours", *J. Theor. Med.* **4**, 21–38.
59. Klarnet, J., Kern, D., Okuno, K., Holt, C., Lilly, F. & Greenberg, P. (1989), "FBL-reactive CD8$^+$ cytotoxic and CD4$^+$ helper T lymphocytes recognize distinct Friend murine leukemia virus-encoded antigens", *J. Exp. Med.* **169**, 457–467.
60. Klein, G., Sjogren, H., Klein, E. & Hellström, K. (1960), "Demonstration of resistance against methylcholanthrene-induced sarcomas in the primary autochtonous host", *Cancer Res.* **20**, 1561–1572.

61. Kuznetsov, V. (1979), Dynamics of cellular immune anti-tumor reactions. I. Synthesis of a multi-level model, in V. Fedorov, ed., "Mathematical Methods in the Theory of Systems", Frunze: Kyrghyz State University, pp. 57–71. (In Russian).

62. Kuznetsov, V. (1991), "A mathematical model for the interaction between cytotoxic lymphocytes and tumor cells. Analysis of the growth, stabilization and regression of the B cell lymphoma in mice chimeric with respect to the major histocompatibility complex", *Biomed. Sci.* **2**(5), 465–476.

63. Kuznetsov, V. & Knott, G. (2001), "Modeling tumor regrowth and immunotherapy", *Math. Comput. Modelling* **33**, 1275–1287.

64. Kuznetsov, V. & Puri, R. (1999), "Kinetic analysis of the high affinity forms of interleukin (IL)-13 receptors: Suppression of IL-13 binding by IL-2 receptor γ chain", *Biophysical J.* **77**, 154–172.

65. Kuznetsov, V., Makalkin, I., Taylor, M. & Perelson, A. (1994), "Nonlinear dynamics of immunogenic tumours: Parameter estimation and global bifurcation analysis", *Bull. Math. Biol.* **56**, 295–321.

66. Kuznetsov, V., Zhivoglyadov, V. & Stepanova, L. (1993), "Kinetic approach and estimation of the parameters of cellular interaction between the immune system and a tumour", *Arch. Immunol. Therap. Exper.* **41**, 21–32.

67. Kwok, C., Cole, S. & Liao, S. (1988), "Uptake kinetics of monoclonal antibodies by human malignant melanoma multicell spheroids", *Cancer Res.* **48**(7), 1856–1863.

68. Kyle, A., Chan, C. & Minchinton, A. (1999), "Characterization of threedimensional tissue cultures using electrical impedance spectroscopy", *Biophys J.* **76**(5), 2640–2648.

69. Lee, J., Moran, J., Fenton, B., Koch, C., Frelinger, J., Keng, P. & Lord, E. (2000a), "Alteration of tumour response to radiation by interleukin-2 gene transfer", *Br J Cancer* **82**(4), 937–944.

70. Lee, L., Hellendall, R., Wang, Y., Haskilla, J., Mukaida, N., Matsushima, K. & Ting, J. (2000b), "IL-8 reduces tumorigenicity of human ovarian cancer in vivo due to neutrophil infiltration", *J Immunol.* **164**(5), 2769–2775.

71. Lefever, R. & Erneaux, T. (1984), On the growth of cellular tissues under constant and uctuating environmental conditions, *in* W. Ross & A. Lawrence, eds, "Nonlinear Electrodynamics in Biological Systems", Plenum Publishing Corporation, pp. 287–305.

72. Lodge, P., Childs, R., Monahan, S. et al. (1999), "Expression and purification of prostate-specific membrane antigen in the baculovirus expression system and recognition by prostate-specific membrane antigen-specific T cells", *J. Immunother.* **22**(4), 346–355.

73. Loeffler, D. & Ratner, S. (1989), "In vivo localization of lymphocytes labeled with low concentrations of HOECHST-33342", *J. Immunol. Meth.* **119**, 95–101.

74. Loeffler, D., Heppner, G. & Lord, E. (1988), "Inuence of hypoxia on T lymphocytes in solid tumours", *Proc. Amer. Assoc. Cancer Res.* **29**, p378.

75. Lord, E. & Burkhardt, G. (1984), "Assessment of in situ host immunity to syngeneic tumours utilizing the multicellular spheroid model", *Cell. Immunol.* **85**, 340–350.

76. Lord, E. & Nardella, G. (1980), "The multicellular tumour spheroid model. 2. characterization of the preliminary allograft response in unsensitized mice", *Transplantation* **29**, 119–124.

77. Lowy, D. & Schiller, J. (1999), "Papillomaviruses: prophylactic vaccine prospects", *Biochim. Biophys. Acta* **1423**(1), M1–M8.

78. Lustgarten, J., Theobald, M., Labadie, C., LaFace, D., Peterson, P., Disis, M., Cheever, M. & Sherman, L. (1997), "Identification of Her-2/neu CTL epitopes using double transgenic mice expressing HLA-A2.1 and human CD8", *Hum. Immunol.* **52**, 109–118.

79. MacSween, R. & Whaley, K., eds (1992), *Muir"s Textbook of Pathology*, 13th edn, Edward Arnold, London.

80. Marušić, M., Ž. Bajzer, Vuk-Pavlović, S. & Freyer, J. (1994a), "Analysis of the growth of multicellular tumour spheroids by mathematical models", *Cell Prolif.* **27**, 73–94.

81. Marušić, M., Ž. Bajzer, Vuk-Pavlović, S. & Freyer, J. (1994b), "Tumour growth in vivo and as multicellular spheroids compared by mathematical models", *Bull. Math. Biol.* **56**, 617–631.

82. Matzavinos, A. & Chaplain, M. (2004), "Travelling wave analysis of a model of the immune response to cancer", *C.R. Biologies* **327**, 995–1008.

83. Matzavinos, A., Chaplain, M. & Kuznetsov, V. (2004), "Mathematical modelling of the spatio-temporal response of cytotoxic T-lymphocytes to a solid tumour", *Mathematical Medicine and Biology: A Journal of the IMA* **21**, 1–34.

84. Matzavinos-Toumasis, A. (2004), Mathematical modelling of the spatiotemporal response of cytotoxic T-lymphocytes to a solid tumour, PhD thesis, University of Dundee.

85. Melief, C., Annels, N., Dunbar, R. et al. (1998), "Protective T-cell responses against tumours", *Res. Immunol.* **149**(9), 877–879.

86. Merlo, G., Siddiqui, J., Cropp, C. et al. (1989), "Frequent alteration of the DF3 tumor-associated antigen gene in primary human breast carcinomas", *Cancer Res.* **49**(24), 6966–6971.

87. Murphy, C. & Newsholme, P. (1999), "Macrophage-mediated lysis of a beta-cell line, tumour necrosis factor-alpha release from bacillus Calmette- Guerin (BCG)-activated murine macrophges and interleukin-8 release from human monocytes are dependent on extracellular glutamine concentration and glutamine metabolism", *Clin Sci (Lond)* **96**(1), 89–97.

88. Murphy, G., Elgamal, A., Su, S. et al. (1998), "Current evaluation of the tissue localization and diagnostic utility of prostate specific membrane antigen", *Cancer* **83**(11), 2259–2269.

89. Naumov, G., MacDonald, I., Weinmeister, P., Kerkvliet, N., Nadkarni, K., Wilson, S., Morris, V., Groom, A. & Chambers, A. (2002), "Persistence of solitary mammary carcinoma cells in a secondary site: a possible contributor to dormancy", *Cancer Res.* **62**(7), 2162–2168.

90. Nesvetov, A. & Zhdanov, A. (1981), "Relationship of the morphology of the immune response and the histological structure of tumours in stomach cancer patients", *Voprosy Onkologii* **27**, 25–31.

91. Nomiyama, H., Hieshima, K., Nakayama, T., Sakaguchi, T., Fujisawa, R., Tanase, S., Nishiura, H., Matsuno, K., Takamori, H., Tabira, Y., Yamamoto, T., Miura, R. & Yoshie, O. (2001), "Human CC chemokine liver-expressed chemokine/CCL16 is a functional ligand for CCR1, CCR2 and CCR5, and constitutively expressed by hepatocytes", *Int Immunol.* **13**(8), 1021–1029.

92. Nosten, F., Luxemburger, C., Kyle, D., Ballou, W., Wittes, J., Wah, E., Chongsuphajaisiddhi, T., Gordon, D., White, N., Sadoff, J. & Heppner, D. (1996),

"Randomized double-bind placebo-controlled trial of SPf66 malaria vaccine in children in northwestern Thailand", *Lancet* **348**, 701–707.

93. Nowak, M. & May, R. (2000), Virus Dynamics: *Mathematical Principles of Immunology and Virology*, Oxford University Press.

94. O'Connell, J., Bennett, M., O'Sullivan, G., Collins, J. & Shanahan, F. (1999), "The Fas counterattack: cancer as a site of immune privilege", *Immunol. Today* **20**(1), 46–50.

95. Old, L., Boyse, E., Clarke, D. & Carswell, E. (1962), "Antigenic properties of chemically induced tumors", *Ann. N. Y. Acad. Sci.* **101**, 80–106.

96. Owen, M. & Sherratt, J. (1997), "Pattern formation and spatio-temporal irregularity in a model for macrophage-tumour interactions", *J. Theor. Biol.* **189**, 63–80.

97. Owen, M. & Sherratt, J. (1998), "Modelling the macrophage invasion of tumours: Effects on growth and composition", *IMA J. Math. Appl. Med. Biol.* **15**, 165–185.

98. Owen, M. & Sherratt, J. (1999), "Mathematical modelling of macrophage dynamics in tumours", *Math. Models Meth. Appl. Sci.* **9**, 513–539.

99. Pardoll, D. (1998), "Cancer vaccines", *Nat. Med.* **4**(5 Suppl), 525–531.

100. Parmiani, G. & Lotze, M., eds (2002), *Tumor Immunology: Molecularly Defined Antigens and Clinical Applications*, Vol. 1 of Tumor Immunology and Immunotherapy Series, Taylor & Francis.

101. Perelson, A. & Weisbuch, G. (1997), "Immunology for physicists", *Reviews of Modern Physics* **69**(4), 1219–1267.

102. Plata, F., Langlade-Demoyen, P., Abastado, J., Berbar, T. & Kourilsky, P. (1987), "Retrovirus antigens recognized by cytolytic T lymphocytes activate tumor rejection in vivo", *Cell* **48**, 231–240.

103. Ponnighaus, J., Fine, P., Sterne, J., Wilson, R., Msosa, E., Gruer, P., Jenkins, P., Lucas, S., Liomba, N. & Bliss, L. (1992), "Efficacy of BCG vaccine against leprosy and tuberculosis in northern Malawi", *Lancet* **339**, 636–639.

104. Prehn, R. (1994), "Stimulatory effects of immune reactions upon the growths of untransplanted tumours", *Cancer Res.* **54**, 908–914.

105. Prehn, R. & Main, J. (1957), "Immunity to methylcholanthrene-induced sarcomas", *J. Natl. Cancer Inst USA* **18**, 759–778.

106. Prigogine, I. & Lefever, R. (1980), "Stability problems in cancer growth and nucleation", *Comp. Biochem. Physiol.* **67**, 389–393.

107. Puri, R. & Siegel, J. (1993), "Interleukin-4 and cancer therapy", *Cancer Invest.* **11**, 473–486.

108. Ratner, S. & Heppner, G. (1986), "Mechanisms of lymphocyte traffic in neoplasia", *Anticancer Res.* **6**, 475–482.

109. Reisenberger, K., Egarter, C., Vogl, S., Sternberger, B., Kiss, H. & Husslein, P. (1996), "The transfer of interleukin-8 across the human placenta perfused in vitro", *Obstet Gynecol.* **87**(4), 613–616.

110. Resnitzky, P., Bustan, A., Peled, A. & Marikovsky, Y. (1988), "Variations in surface charge distribution of leukemic and nonleukemic transformed cells", *Leuk. Res.* **12**, 315–320.

111. Ressing, M., de Hong, J., Brandt, R. et al. (1999), "Differential binding of viral peptides to HLA-A2 alleles. Implications for human papillomavirus type 16 E7 peptide-based vaccination against cervical carcinoma", *Eur. J. Immunol.* **29**(4), 1292–1303.

112. Retsky, M., Wardwell, R., Swartzendruber, D. & Headley, D. (1987), "Prospective computerized simulation of breast-cancer – comparison of computer-predictions with 9 sets of biological and clinical data", *Cancer Res.* **47**, 4982–4987.

113. Robbins, P. (2002), Immune Recognition of Cancer – Tumor Antigens, in G. Parmiani & M. Lotze, eds, "Tumor Immunology: Molecularly Defined Antigens and Clinical Applications", Taylor & Francis, pp. 11–46.

114. Robbins, P., el Gamil, M., Li, Y., Topalian, S., Rivoltini, L., Sakaguchi, B., Appella, E., Kawakami, Y. & Rosenberg, S. (1995), "Cloning of a new gene encoding an antigen recognized by melanoma-specific HLA-A24- restricted tumor-infiltrating lymphocytes", *J. Immunol.* **154**, 5944–5950.

115. Rosenberg, S. (1999), "A new era for cancer immunotherapy based on the genes that encode cancer antigens", *Immunity* **10**(3), 281–287.

116. Rothstein, T., Mage, M., Mond, J. & McHugh, L. (1978), "Guinea pig antiserum to mouse cytotoxic T-lymphocytes and their precursor", *J. Immunol.* **120**, 209–215.

117. Ruiz-Cabello, F. & Garrido, F. (2002), Tumor evasion of immune system, in G. Parmiani & M. T. Lotze, eds, "Tumor Immunology: Molecularly Defined Antigens and Clinical Applications", Taylor & Francis, pp. 177–203.

118. Salgaller, M. (1997), "Monitoring of cancer patients undergoing active or passive immunotherapy", *J. Immunother.* **20**, 1–14.

119. Satyamoorthy, K., Li, G., Vaidya, B., Kalabis, J. & Herlyn, M. (2002), "Insulin-like growth factor-I-induced migration of melanoma cells is mediated by interleukin-8 induction", *Cell Growth Differ.* **13**(2), 87–93.

120. Schlom, J. & Abrams, S. (2000), Tumor Immunology, in R. Bast, D. Kufe, R. Pollock, R. Weichselbaum, J. Holland & E. F. III, eds, "Cancer Medicine", 5th edn, BC Decker, pp. 153–167.

121. Schwartzentruber, D., S.L.Topalian, Mancini, M. & Rosenberg, S. (1991), "Specific release of granulocyte-macrophage colony-stimulating factor, tumour necrosis factor-α and IFN-γ by human tumour-infiltrating lymphocytes after autologous tumour stimulation", *J. Immunol.* **146**, 3674–3681.

122. Shampine, L. & Reichelt, M. (1997), "The MATLAB ODE suite", *SIAM J. Sci. Comput.* **18**(1), 1–22.

123. Sherbet, G. (1989), "Membrane uidity and cancer metastasis", *Expl. Cell. Biol.* **57**, 198–205.

124. Sherratt, J., Lewis, M. & Fowler, A. (1995), "Ecological chaos in the wake of invasion", *Proc. Natl. Acad. Sci. USA* **92**, 2524–2528.

125. Siu, H., Vitetta, E., May, R. & Uhr, J. (1986), "Tumour dormancy. regression of BCL tumour and induction of a dormant tumour state in mice chimeric at the major histocompatibility complex", *J. Immunol.* **137**, 1376–1382.

126. Sordat, B., MacDonald, H. & Lees, R. (1980), "The multicellular spheroid as a model tumour allograft. 3. morphological and kinetic analysis of spheroid infiltration and destruction", *Transplantation* **29**, 103–112.

127. Sozzani, S., Luini, W., Molino, M., Jilek, P., Bottazzi, B. & Cerletti, C. (1991), "The signal transduction pathway involved in the migration induced by a monocyte chemotactic cytokine", *J Immunol.* **147**, 2215–2221.

128. Sutherland, R. (1988), "Cell and environment interactions in tumor microregions: the multicell spheroid model", *Science* **240**(4849), 177–184.

129. Suzuki, Y., Liu, C., Chen, L., Bennathan, D. & Wheelock, E. (1987), "Immune regulation of the L5178Y murine tumour dormant state. 2. interferon gamma requires tumour necrosis factor to restrain tumour cell growth in peritoneal cell cultures from tumour dormant mice", *J. Immunol.* **139**, 3146–3152.

130. Tanaka, Y., Tevethia, M., Kalderon, D., Smith, A. & Tevethia, S. (1988), "Clustering of antigenic sites recognized by cytotoxic T lymphocyte clones in the amino terminal half of SV40 antigen", *Virology* **162**, 427–436.

131. Tjoa, B., Elgamal, A. & Murphy, G. (1999), "Vaccine therapy for prostate cancer", *Urol. Clin. North Am.* **26**(2), 365–374.

132. Trinchieri, G., Aden, D. & Knowles, B. (1976), "Cell mediated cytotoxicity to SV40 specific tumor associated antigens", *Nature* **261**, 312–314.

133. Udagawa, T., Fernandez, A., Achilles, E., Folkman, J. & D"Amato, R. (2002), "Persistence of microscopic human cancers in mice: alterations in the angiogenic balance accompanies loss of tumor dormancy", *FASEB J.* **16**(11), 1361–1370.

134. Uhr, J. & Marches, R. (2001), "Dormancy in a model of murine B cell lymphoma", *Seminars in Cancer Biology* **11**, 277–283.

135. Uhr, J., Tucker, T., May, R., H.Siu & Vitetta, E. (1991), "Cancer dormancy: Studies of the murine BCL lymphoma", *Cancer Res.* (Suppl.) **51**, 5045s–5053s.

136. van Blitterswijk, W. (1988), "Structural basis and physiological control of membrane uidity in normal and tumor cells", *Subcellular Biochemistry* **13**, 393–413.

137. van Driel, W., Kenter, G., Fleuren, G. et al. (1999), "Immunotherapeutic strategies for cervical squamous carcinoma", *Hematol. Oncol. Clin. North Am.* **13**(1), 259–273.

138. Vaughan, H. (1914), "Cancer vaccine and anti-cancer globulin as an aid in the surgical treatment of malignancy", *J. Am. Med. Ass.* **63**, 1258–1268.

139. Webb, S., Sherratt, J. & Fish, R. (2002), "Cells behaving badly: a theoretical model for the Fas/FasL system in tumour immunology", *Mathematical Biosciences* **179**, 113–129.

140. Wheelock, E.,Weinhold, K. & Levich, J. (1981), "The tumour dormant state", *Adv. Cancer Res.* **34**, 107–140.

141. Wilson, K. & Lord, E. (1987), "Specific (EMT6) and non-specific (WEHI-164) cytolytic activity by host cells infiltrating tumour spheroids", *Brit. J. Cancer* **55**, 141–146.

Control Theory Approach to Cancer Chemotherapy: Benefiting from Phase Dependence and Overcoming Drug Resistance

Marek Kimmel[1,2] and Andrzej Swierniak[1]

[1] Department of Automatic Control, Silesian University of Technology, 44-101 Gliwice, Akademicka 16, Poland
[2] Department of Statistics, Rice University, Houston, Texas, USA

Abstract. Two major obstacles against successful chemotherapy of cancer are (1) cell-cycle-phase dependence of treatment, and (2) emergence of resistance of cancer cells to cytotoxic agents. One way to understand and overcome these two problems is to apply optimal control theory to mathematical models of cell cycle dynamics. These models should include division of the cell cycle into subphases and/or the mechanisms of drug resistance. We review our results in mathematical modeling and control of the cell cycle and of the mechanisms of gene amplification (related to drug resistance), and estimation of parameters of the constructed models.

1 Introduction

In this paper we are concerned with three issues:

1. The inner structure of the cell cycle and the cell-cycle-phase specificity of some chemotherapy agents.
2. The dynamics of emergence of resistance of cancer cells to chemotherapy, as understood based on recent progress in molecular biology.
3. Estimation of quantitative parameters of the cell cycle, drug action and cell mutation to resistance.

The main purpose of the paper is to outline our own views on the issues involved. The paper is in large part a critical survey of published work by us and others. Wherever appropriate, we give credit to others, without attempts at an exhaustive review.

The philosophy of this paper is related to our professional experience. The first author has spent almost ten years in a cancer research institute trying to develop models of the cell cycle for the purpose of estimation of cell-cycle-phase specific action of anticancer drugs. In addition, he investigated gene amplification as the mechanism of resistance of cancer cells. The other author has

M. Kimmel and A. Swierniak: *Control Theory Approach to Cancer Chemotherapy: Benefiting from Phase Dependence and Overcoming Drug Resistance*, Lect. Notes Math. **1872**, 185–221 (2006)
www.springerlink.com

been involved for two decades in attempts to develop a satisfactory theory of optimal control of bilinear systems resulting from a description of chemotherapy action using ordinary differential equations. The cell-cycle-phase specificity is essential for the initial period of chemotherapy, when at issue is the most efficient reduction of the cancer burden. This seems to be of practical importance mainly in nonsurgical cancers such as for example leukemias. Emergence of clones of cancer cells resistant to chemotherapy is important in treatment and prevention of systemic spread of disease. This comprises potential treatment of metastasis and all variants of adjuvant chemotherapy.

Mathematical modeling of cancer chemotherapy has had more than four decades of history. It has contributed to the development of ideas of chemotherapy scheduling, multidrug protocols, and recruitment. It has also helped in the refinement of mathematical tools of control theory applied to the dynamics of cell populations [39]. However, regarding practical results it has been, with minor exceptions, a failure. The reasons for that failure are not always clearly perceived. They stem from the direction of both biomedicine and mathematics: important biological processes are ignored and crucial parameters are not known, but also the mathematical intricacy of the models is not appreciated.

In this paper, we would like to outline several directions of research which may play a role in improving the situation and realizing the obvious potential existing in the mathematical approach. Because of recent progress in methods of monitoring cancer cell populations, new insights and more precise measurements became possible. This, together with a progress in mathematical tools, has renewed hopes for improving chemotherapy protocols.

Cell-cycle-phase specificity of some cytotoxic drugs is important since it makes sense to apply anticancer drugs when cells gather in the sensitive phases of the cell cycle. It can be approached by considering dissection of the cell cycle into an increasing number of disjoint compartments, with drug action limited to only some of them. We provide a classification of several simplest models of this kind. Mathematical problems encountered include singularity and non-uniqueness of solutions of the optimization problems.

The emergence of resistance to chemotherapy has been first considered in a point mutation model of Coldman and Goldie (e.g. [30, 47]) and then in the framework of gene amplification by Agur and Harnevo (e.g. [52–54]). The main idea is that there exist spontaneous or induced mutations of cancer cells towards drug resistance and that the scheduling of treatment should anticipate these mutations. The point mutation model can be translated into simple recommendations, which have even been recently tested in clinical trials. The gene amplification model was extensively simulated and also resulted in recommendations for optimized therapy. We present a model of chemotherapy based on a stochastic approach to evolution of cancer cells. Asymptotic analysis of this model results in some understanding of its dynamics. This, in our opinion, is the first step towards a more rigorous mathematical treatment of the dynamics of drug resistance and/or metastasis. Optimization of the chemotherapy in this case may be viewed as the progress in creating chemotherapy resistant to drug resistance.

There is no doubt that the parameters of spontaneous cancer cells populations existing in vivo in humans differ considerably from those of the test tube "transformed" cells and from those of the induced animal tumors. However, much information regarding cell-cycle-phase specificity of anticancer agents has been obtained using in vitro experimental models. We present some of these results which do not seem to be sufficiently well known. We also discuss some approaches to estimation of cell cycle parameters of human tumors. Finally we discuss estimation of the rates of mutations leading to drug resistance.

2 Modeling the Cell Cycle

The cell cycle is composed of a sequence of phases traversed by each cell from its birth to division. These phases are: G_1, or the growth phase; S, or the DNA synthesis phase; G_2, or the preparation for division phase; and M, or the division phase. After division, the two daughter cells usually re-enter G_1. It may however happen that one or both daughters deviate from this path and become dormant or resting, or in other words, they enter the quiescent G_0 phase. From there after a variable and usually rather long time cells may reenter the cell cycle in G_1 [9].

This idealized scheme is confounded in solid tumors by the existence of a geometric gradient of availability of oxygen and nutrients. This causes a stratification in viability of cells: usually, cycling cells are located near the surface or near blood vessels, further layers are occupied by dormant cells, while the deepest regions form a necrotic core. This may lead to self-limiting growth phenomena, which may be described by biologically based nonlinear models including Pearl-Verhulst, Cox-Woodbury-Meyers, Michaelis-Menton equations (see e.g. [90, 113–116]), or Gompertz-type equations [50, 108, 110, 117, 142, 148]. It is interesting to note that Gompertz proposed his model for demographic purposes [48], and its biological meaning is difficult to justify, but the Gompertzian growth is a good approximation of a number of experimental data for tumor growth. We do not consider this structure in our models. Instead we build a set of models of cell cycle kinetics composed of compartments (see e.g. [59, 133]) each of them containing a phase or a cluster of phases.

The transit times through all the phases of the cell cycle are variable, particularly in malignant cells. Usually it is assumed that most of this variability is concentrated in the G_1 phase (and in G_0 whenever it exists). The simplest models arise if the transit times through each compartment are assumed exponentially distributed.

Denote by $N_i(t)$ the average number of cells in the i-th compartment at time t, and by $x_i^+(t)$ and $x_i^-(t)$, the average flow rates of cells into and out of this compartment, respectively. Then,

$$\dot{N}_i(t) = x_i^+(t) - x_i^-(t) , \tag{1}$$

and

$$x_i^-(t) = a_i N_i(t) , \tag{2}$$

where a_i is the parameter of the exponential distribution, equal to the inverse of the average transit time. If the preceding compartment is numbered $i-1$, then

$$\dot{N}_i(t) = -a_i N_i(t) + a_{i-1} N_{i-1}(t) . \tag{3}$$

for $i = 2, 3, \ldots, n$ where n is a number of compartments. The boundary condition for the obtained set of equations is given by:

$$\dot{N}_1(t) = -a_1 N_1(t) + 2a_n N_n(t) . \tag{4}$$

Therefore, under the exponentiality assumption, the unperturbed dynamics of cell cycle, i.e. the number of cells in various cell cycle compartments versus time, in the absence of external stimuli, is expressed by a system of ordinary linear differential equations. We consider three types of perturbations of the cell cycle [118, 128]:

Cell Killing. At time t, only a fraction $u(t)$ of the outflux from compartment i contains viable cells ($0 \leq u(t) \leq 1$). The remaining cells are dead and no longer considered part of the system.

$$\dot{N}_i(t) = -a_i N_i(t) + a_{i-1} N_{i-1}(t) , \tag{5}$$
$$\dot{N}_{i+1}(t) = -a_{i+1} N_{i+1}(t) + u(t) a_i N_i(t) , \tag{6}$$

The reproductively dead cells may however continue to progress through the cycle for some time, thus confounding estimates of cell proliferation.

Cell Arrest. At time t, the outflux from compartment i is reduced to a fraction $v(t)$ of the normal value ($0 < v_m \leq v(t) \leq 1$) . The remaining cells are arrested in compartment i.

$$\dot{N}_i(t) = -v(t) a_i N_i(t) + a_{i-1} N_{i-1}(t) , \tag{7}$$
$$\dot{N}_{i+1}(t) = -a_{i+1} N_{i+1}(t) + v(t) a_i N_i(t) . \tag{8}$$

The complete arrest is not possible, and it is why v_m is always strictly positive.

Alteration of the Transit Time. The parameter of the exponential distribution of the transit time through compartment i is changed by factor $y(t) > 0$. Depending on whether $y(t)$ is less or greater than 1, this is equivalent to respectively extending or reducing the mean transit time. In the latter case it is used in the so called recruitment of dormant cells to the proliferation cycle. The mathematical description has the form:

$$\dot{N}_i(t) = -y(t)a_i N_i(t) + a_{i-1} N_{i-1}(t) , \tag{9}$$

$$\dot{N}_{i+1}(t) = -a_{i+1} N_{i+1}(t) + y(t)a_i N_i(t) , \tag{10}$$

Formally, these equations are identical as those describing cell arrest. This effect is caused by the exponentiality assumption.

Since our models describe an average behaviour of considered subpopulations the compartments which they represent are sometimes called "probabilistic" or "statistical" ones. "Deterministic" description of the continuously dividing population should employ partial differential equations (for example of von Förster [43]), integro-differential or integral equations (for example [65]). In this case the one independent variable represents the chronological time while the other age or size. In our models the age is simply discretized and the dynamics from one stage to the other is averaged. The combined approach is to use both "probabilistic" and "deterministic" compartments to model the cell cycle as for example in [34, 35].

The first class of drug actions is represented by G_2/M specific agents, which include the so-called spindle poisons like Vincristine, Vinblastine or Bleomycin which destroy a mitotic spindle [25] and Taxol [42] or 5-Fluorouracil [26] affecting mainly cells during their division. Killing agents also include S specific drugs like Cyclophosphamide [42] or Methotrexate – MTX [96] acting mainly in the DNA replication phase, Cytosine Arabinoside – Ara-C, rapidly killing cells in phase S through inhibition of DNA polymerase by competition with deoxycytosine triphosphate [32]. Among the blocking drugs used to arrest the cells immediately before or during DNA synthesis we can mention antibiotics like Adriamycin, Daunomycin, Dexorubin, Idarudicin which cause the progression blockage on the border between the phases G_1 and S by interfering with the formator of the polymerase complex or by hindering the separation of the two polynucleotide strands in the double helix [4]. Another blocking agent is Hydroxyurea – HU [35, 84] which is found to synchronize cells by causing brief and invisible inhibition of DNA synthesis in the phase S and holding cells in G_1. The recruitment action was demonstrated [5] for Granulocyte Colony Stimulating Factors – G-CSF, Granulocyte Macrophage Colony Stimulating Factors – GM-CSF, Interleukin-3 – Il-3, specially when combined with Human Cloned Stem Cell Factor – SCF.

This classification of anticancer agents is not quite sharp and there is some controversy in the literature concerning both the site and the role of action of some drugs. For example, although mostly active in specific phases Cyclophosphamide and 5-Fluorouracil kill cells also in other phases of the proliferation cycle that enables to encounter them to cycle specific agents [21, 25]. On the other hand some antimitotic agents like curacin A [77] act by increasing the S phase transition time (blocking) and decreasing the M phase transition time.

Killing agents which we consider in our model are applied in the G_2/M phase which makes sense from a biological standpoint for a couple of reasons. First, in mitosis M the cell becomes very thin and porous. Hence, the cell is more vulnerable to an attack while there will be a minimal effect on the normal cells. Second, chemotherapy during mitosis will prevent the creation of daughter cells.

While the killing agent is the only control considered in the two-compartment model below, in the three-compartment model in addition a blocking agent is considered which slows down the development of cells in the synthesis phase S and then releases them at the moment when another G_2/M specific anticancer drug has maximum killing potential (so-called synchroniza-tion [22]). This strategy may have the additional advantage of protecting the normal cells which would be less exposed to the second agent (e.g. due to less dispersion and faster transit through G_2/M) [2, 34]. This cell cycle model includes separate compartments for the G_0/G_1, S and G_2/M phases.

One of the major problems in chemotherapy of some leukemias is consti-tuted by the large residuum of dormant G_0 cells which are not sensitive to most cytotoxic agents [26, 57, 83]. Similar findings for breast and overian cancers were reported, e.g. in [28, 42]. As indicated by these authors the insensitivity of dormant cells to the majority of anticancer drugs and percentage of tumor mass resting is a fact which, if ignored, leads not only to clinical problems but also to some erronomous theoretical considerations. Experiments with Ara-C [32], indicated that while double injected during cell cycle or combined with Adriamycin or anthracyclines led to serious reduction of leukemic burden without an evident increase of negative effect on normal tissues. This thera-peutic gain was attributed to the specific recruitment inducing effect of Ara-C on leukemic cells in the dormant phase It became possible to efficiently recruit quiescent cells into the cycle using cytokines [132] (substances playing a role in the regulation of normal hemopoiesis) like G-CSF, GM-CSF, and especially Il-3 combined with SCF. Then, a cytotoxic agent like Ara-C or anthracyclines may be used. The other three compartment model below uses separate com-partments for the G_0, G_1 and $S+G_2/M$ phases and includes such a recruiting agent. Moreover, it enables also analysis of the alteration of the transit time through G_0 phase due to the feedback mechanism that recruits the cells into the cycle when chemotherapy is applied. In a similar way we may model other types of manipulation of the cell cycle as for example the use of triterpenoids to inhibit proliferation and induce differentiation and apoptosis in leukemic cells [75].

The important assumption which is satisfied in all our models is that the control systems which they represent are *internally positive* [60] i.e.:

(+) The first orthant of the control system is positively invariant, that is for any admissible control and or positive initial states, the state remains positive for all times $t > 0$.

Thus the obvious modelling state-space constraints that the state is positive, need not be included in our model explicitly and the analysis simplifies. A simple sufficient condition for (+) to hold (for example, see [60]) is that:

(M) all the system matrices for all admissible controls are so-called M-matrices, i.e. have negative diagonal entries, but non-negative off-diagonal entries.

This condition is natural and will be satisfied for any compartmental model whose dynamics is given by balance equations where the diagonal entries correspond to the outflows from the i-th compartments and the off-diagonal entries represent the inflows from the i-th into the j-th compartment, $i \neq j$. It is satisfied for each of the models described here. More generally, if condition $(+)$ were violated, this is a strong indication that the modelling is inconsistent.

3 Control Problems with Cell-Cycle-Phase Dependence

The classical control theoretic design problem may be stated as follows. Let the dynamic properties of a system be described by its state, and the external actions i.e. control and disturbances be given by input variables. Moreover assume that we are given a target set of required system states or outputs. Find control actions which enable reachability of the desirable region. If we are able to describe a disease by a finite number of dynamically changing parameters we are also able to formulate the control problem in the sense mentioned above. In the models considered here the problem of finding an optimal cancer chemotherapy protocol is formulated as an optimal control problem over a finite time-interval, the fixed therapy horizon. The state variable is given by the average number of cancer cells and the control is the effect of the drug dosages on the respective subpopulation. The goal is to maximize the number of cancer cells which the agent kills, respectively minimize the number of cancer cells at the end of the therapy session, while keeping the toxicity to the normal tissues acceptable. The latter aspect is modelled implicitly by including an integral of the control over the therapy interval in the objective so that minimizing controls will have to balance the amount of drugs given with the conflicting objective to kill cancer cells.

From the first attempts at cell-cycle-phase dependent chemotherapy, one of the central ideas was that of *synchronization* [79]. In one version, the concept includes using an agent to arrest cells before they enter a sensitive phase. After enough of them accumulate, they are released into the sensitive phase and then targeted by a killing agent. This tactic is employed in the first of our three-compartments models. However, synchronization may be achieved, at least in theory, using only one agent, by periodic administration (see e.g. [35, 140]). At appropriate frequency and dosage, maximum efficiency would be achieved (the *resonance*). This tactic may be combined with attempts at sparing the normal cells by taking advantage of the difference in cell cycle duration of cancer and normal cells. This problem was considered in a number of papers by Agur and coworkers (e.g. [2, 3, 29]). Their line of reasoning is based on the so called Z-method in which the crucial parameter is the elimination coefficient Z measuring the treatment efficacy defined by

$$Z = 1 - \frac{T_m}{T_h} \, , \tag{11}$$

where T_m is the elimination time of the malignant population and T_h is the one of critical host population. Agur et al. [3] find that treatment efficacy is a nonmonotonic function of the relation between the cell generation time and the period of drug administration with maxima occurring when the critical host cell cycle length is a multiple of the chemotherapeutic period. The results in the papers imply that short drug-pulses at appropriate intervals may be more efficient than a drug administered at arbitrary intervals or a continuous slowly released drug. Under the condition that the cell cycle parameters of malignant cells have a relatively large variation, the drug protocol could be determined by the host temporal parameters alone and should reduce cytotoxity even in the case of similar mean cell cycle times for cancer and normal tissues.

3.1 Single Compartment, Single Killing Agent

In the simplest model it is assumed that the cytotoxic agent is not cell-cycle-phase specific [72]. Therefore, whole cell cycle is modeled as a single compartment. The corresponding single differential equation has the form,

$$\dot{N}(t) = -aN(t) + 2u(t)aN(t), \; N(0) = N_0, 0 \leq u(t) \leq 1 \, . \tag{12}$$

Control variable $u(t)$ assumes values $u(t) = 1$ when the drug is not administered, $u(t) = 0$ when the maximum dose is used, and $0 < u(t) < 1$ in all other cases.

This bilinear model is used to find the optimal control which minimizes the performance index,

$$J = rN(T) + \int_0^T [1 - u(t)] \, dt \, . \tag{13}$$

In biological terms, the effect of the optimal control is minimization of the number of cancer cells at the end of the assumed therapy interval [0, T], combined with minimization of the cumulative negative effects of the drug upon the normal tissues; r is a weighing coefficient.

This optimization problem is mathematically so simple that it can be explicitly solved. Substituting the solution of equation (12) into (13) yields,

$$J = rN(T) + \frac{1}{2}T + \frac{1}{2a} \ln \left[\frac{N_0}{N(T)} \right] \, , \tag{14}$$

and its minimum value is obtained for

$$N(T) = \frac{1}{2ar} \, , \tag{15}$$

if the following inequality is satisfied,

$$0 \leq \frac{1}{2}T - \frac{1}{2a}\ln(2arN_0) = T_1 \leq T . \tag{16}$$

This inequality results from the constraints imposed on the control variable. In this case, any control satisfying the relationship,

$$\int_0^T u(t)dt = \frac{1}{2}T + \frac{1}{2a}\ln\left[\frac{N_0}{N(T)}\right] = \frac{1}{2}T - \frac{1}{2a}\ln(2arN_0) , \tag{17}$$

is optimal.

In [121] we explain the nonuniqueness of the solution by its total singularity. Moreover in [120] we have shown that extensions of the first order model assuming Bellman's model of pharmacokinetics for anticancer drug (see [16, 31]), or simultaneously considering the drug effect on cancer and normal proliferating cells, do not enable us to avoid the singularity of optimal control.

3.2 Two Compartments, Single G_2M – Specific Killing Agent

This is probably the simplest situation in which it is possible to contemplate the effects of phase specificity [119, 123]. Compartment 1 consists of the G_1 and S phases and compartment 2 of the G_2 and M phases. The corresponding system of two differential equations has the form,

$$\begin{aligned}
\dot{N}_1(t) &= -a_1 N_1(t) + 2ua_2 N_2(t), \quad N_1(0) = N_{10} > 0 , \\
\dot{N}_2(t) &= -a_2 N_2(t) + a_1 N_1(t), \qquad N_2(0) = N_{20} > 0 .
\end{aligned} \tag{18}$$

The performance index has the form analogous to (14),

$$J = \sum_{i=1}^{2} r_i N_i(T) + \int_0^T [1 - u(t)]\, dt , \tag{19}$$

and its interpretation is identical as before.

If the optimal control is of the bang-bang type, it can be found from the maximum principle [100] by minimizing the so called hamiltonian function:

$$H = p_1(-a_1 N_1 + 2ua_2 N_2) + p_2(-a_2 N_2 + a_1 N_1) + 1 - u , \tag{20}$$

that results in:

$$u(t) = \begin{cases} 0; & 2a_2 N_2(t)p_1(t) > 1 , \\ 1; & 2a_2 N_2(t)p_1(t) < 1 , \end{cases} \tag{21}$$

where $p = (p_1, p_2)^T$ is the costate vector defined by the conjugate equations,

$$\begin{aligned}
\dot{p}_1(t) &= a_1(p_1(t) - p_2(t)), & p_1(T) &= r_1 , \\
\dot{p}_2(t) &= a_2(p_2(t) - 2p_1(t)u(t)), & p_2(T) &= r_2 ,
\end{aligned} \tag{22}$$

Since the control system satisfies condition (M), then it follows from the adjoint equation that for any admissible control the first orthant in costate-space is negatively invariant under the flow of the adjoint system, i.e. if $p_i(T) > 0$ for all $i = 1, 2$, then $p_i(t) > 0$ for all times $t \leq T$. In this case, since $N(0)$ and $p(T)$ have positive components, it follows that all states N_i and costates p_i are positive over $[0, T]$.

The case $2a_1 N_2 p_2 = 1$ leads to the singular control problems which cannot be excluded using only the first order necessary conditions .

The standard method to solve the problem is to find a numerical solution of the two point boundary value problem (TPBVP) which may be performed using Mohler's STVM [87, 103], semianalytical shooting algorithm [123] or gradient type methods [37, 38]. Among the other methods used to solve optimal control problems arising in chemotherapy scheduling we should mention control parametrization techniques developped by Teo and Martin (see [85, 86]). Numerical studies do not exhibit the whole complexity of the problem. By finding invariance properties of the solutions to the TPBVP on the torus and formulating a special symmetry relation we have been able to classify [126] all the solutions to TPBVP problems. The analysis has indicated the irregularity of the optimal control problem [127], arising from multiplicity of solutions [98], existence of periodic trajectories [122] and existence of singular solutions [121]. The classification of complete trajectories enables to avoid a major disadvantage of a penalty method which has been used in formulation of the performance index i.e. no systematic way of choosing the value of weighting vector. Since final values of the costate vector are weighting parameters in the performance index the analysis of solutions for all possible boundary conditions allows for consideration of their sensitivity to the value chosen for r. The regions of r for which the multiple solutions of the optimal control problem may appear can also be easily assigned [128]. To avoid them, additional constraints may be imposed for the process of reducing the tumor burden. One of the reasonable requirements is that the tumor-population decreases faster than a given rate as proposed by Sundareshan and Fundakowski [111, 112] for multicompartmental models. This constraint may be for example satisfied for the periodic solutions [124].

Moreover recently singularity of optimal arcs was excluded with the use of high-order necessary conditions for optimality and sufficient conditions for optimal bang-bang strategies were found which enable to determine whether controls found by the use of Pontryagin maximum principle are at least locally optimal [80]. More precisely singular controls are calculated by differentiating the switching function in time until the control variable explicitly appears in the derivative, then finding the control which makes it equal to 0. For a single-input system which is linear in the control it is known [78] that the order of this derivative must be even, say $2k$, and k is called the order of the singular arc on the interval I. It is a necessary condition for optimality of a singular arc of order k, the so-called generalized Legendre-Clebsch condition [78], that

$$(-1)^k \frac{\partial}{\partial u} \frac{d^{2k}}{dt^{2k}} \frac{\partial H}{\partial u} \geq 0 . \tag{23}$$

Note that the term $\frac{\partial H}{\partial u}$ in (23) represents the switching function for the problem. This framework directly applies to the 2-compartment model which has a scalar control. Elementary and direct calculations show that in this case singular arcs are of order 1 and that

$$\frac{\partial}{\partial u} \frac{d^2}{dt^2} \frac{\partial H}{\partial u} = 4a_1 a_2 > 0 \tag{24}$$

violating the Legendre-Clebsch condition. To develop sufficient conditions for local optimality field-theoretic concepts have been used. Essentially, if the flow of the system is a diffeomorphism away from the switching surfaces and if it crosses the switching surfaces transversally, then using the method of characteristics a differentiable solution to the Hamilton-Jacobi-Bellman equation can be constructed [92]. This then implies optimality of the flow. The transversality condition:

$$\left| \frac{d}{dt} (N_2(t_k) p_1(t_k)) \right| + 2a_2 N_2(t_k) S(t_k) > 0 \tag{25}$$

should be checked at each crossing of the switching surfaces at time t_k To find $S(t_k)$ a matrix discrete Riccati type equation should be solved iteratively for the moments of control switchings. The sufficient conditions lead to yet another numerical algorithm for the optimal protocols design based on backward integration of the combined flow along the characteristics.

3.3 Three Compartments, Cell Arrest in S and Killing in G_2M

One of the conceivable strategies of protocol optimization, exploiting drug specificity, is to arrest cancer cells in the S phase [22, 49], and then release them at the moment when another G_2M specific anticancer drug has the maximum killing potential. This strategy may have the additional advantage of protecting the normal cells which would be less exposed to the second agent (e.g. due to less dispersion and faster transit through G_2M). The cell cycle model includes separate compartments for the G_1, S and G_2M phases [119, 123].

The control problem is to find $u(t) \in [0,1]$ and $v(t) \in [v_m, 1]$ such that

$$\begin{aligned}
\dot{N}_1(t) &= -a_1 N_1(t) + 2u(t)a_3 N_3(t), & N_1(0) &= N_{10} > 0 , \\
\dot{N}_2(t) &= -v(t)a_2 N_2(t) + a_1 N_1(t), & N_2(0) &= N_{20} > 0 . \\
\dot{N}_3(t) &= -a_3 N_3(t) + v(t)a_2 N_2(t), & N_3(0) &= N_{30} > 0 .
\end{aligned} \tag{26}$$

and the index

$$J = \sum_{i=1}^{3} r_i N_i(T) + \int_0^T [1 - u(t)]\, dt , \tag{27}$$

is minimized. The bang-bang solution found from the maximum principle has the following form

$$u(t) = \begin{cases} 0; & 2a_3 N_3(t) p_1(t) > 1 \\ 1; & 2a_3 N_3(t) p_1(t) < 1 \end{cases} \tag{28}$$

$$v(t) = \begin{cases} v_m; & p_2(t) < p_3(t) \\ 1; & p_2(t) > p_3(t) \end{cases} \tag{29}$$

where the costate vector satisfies the following set of equations,

$$\begin{aligned} \dot{p}_1(t) &= a_1(p_1(t) - p_2(t)), & p_1(T) &= r_1 , \\ \dot{p}_2(t) &= a_2(p_2(t) - p_3(t))v(t), & p_2(T) &= r_2 , \\ \dot{p}_3(t) &= a_3(p_3(t) - 2p_1(t))u(t), & p_3(T) &= r_3 , \end{aligned} \tag{30}$$

The arising TPBVP may be once more treated numerically [38] by the gradient method in the way similar as for two-compartmental models. Analytical treatment becomes much more complicated since the problem could not be projected into the plane. But also in this case it is possible to eliminate singular controls as not optimal and formulate sufficient conditions for local optimality of bang-bang strategies [81]. In this case the generalized Legendre-Clebsch condition (23) still applies to the first control u if we freeze the second control v. Assuming v is constant, it can be shown that a singular control u must be of order 2, but again (23) is violated. Direct, but longer calculations verify that

$$\frac{\partial}{\partial u} \frac{d^4}{dt^4} \frac{\partial H}{\partial u} = -12a_1 a_2 a_3^2 v(a_1 + a_2 v) p_1(t) N_2(t) < 0 . \tag{31}$$

Furthermore, if the control v is singular on an interval I, then it can easily be seen that u also must be singular on I. In this case it is a necessary condition for optimality, the so-called Goh condition [78], that on I we have

$$\frac{\partial}{\partial v} \frac{d}{dt} \frac{\partial H}{\partial u} \equiv 0 . \tag{32}$$

However, a direct calculation gives

$$\frac{\partial}{\partial v} \frac{d}{dt} \frac{\partial H}{\partial u} = 2a_2 a_3 p_1(t) N_2(t) > 0 \tag{33}$$

violating the Goh-condition . Derivation of the sufficient conditions of local optimality for bang-bang strategies follows the similar line as for the two-compartment model. Of course transversality conditions should be checked for both switching controls and to do this we are led to a system of discrete Riccati equations which should be iterated in all considered switching moments for both control variables. An additional assumption, formulated only for technical reasons, is that there are no simultaneous switchings for both controls.

3.4 Three Compartments, Cell Recruitment from G_0 and Killing in G_2M

One of the major problems in chemotherapy of some leukemias is constituted by the large residuum of dormant G_0 cells which are not sensitive to most cytotoxic agents. It became recently possible to efficiently recruit these cells into the cycle using cytokines [5, 132], substances playing a role in the regulation of normal hemopoiesis. Then, a cytotoxic agent may be used. To model such a system, we use separate compartments for the G_0, G_1 and S + G_2M phases, numbered 0, 1 and 2 [128].

The control problem is to find $u(t) \in [0, 1]$ and $y(t) \in [1, y_m]$ such that

$$
\begin{aligned}
\dot{N}_0(t) &= -ya_0 N_0(t) + 2b_0 u(t) a_2 N_2(t), & N_0(0) &= N_{00} > 0 , \\
\dot{N}_1(t) &= -a_1 N_1(t) + ya_0 N_0(t) + 2b_1 u(t) a_2 N_2(t), & N_1(0) &= N_{10} > 0 . \\
\dot{N}_2(t) &= -a_2 N_2(t) + a_1 N_1(t), & N_2(0) &= N_{20} > 0 .
\end{aligned}
\tag{34}
$$

where b_0 and b_1 are the probabilities of the daughter cell entering after division G_0 and G_1, respectively. The index to be minimized is

$$
J = \sum_{i=0}^{2} r_i N_i(T) + \int_0^T [1 - u(t)] \, dt .
\tag{35}
$$

An interesting special case is $N_{00} > 0, N_{10} = N_{20} = 0, r_0 > 0, r_1 = r_2 = 0$, i.e. all the cells concentrated in G_0 at the onset of the therapy, the principal purpose of the therapy being minimization of their eventual number based on presumption that this will yield the eventual demise of the whole population.

The bang-bang solution found using the maximum principle has the following form

$$
u(t) = \begin{cases} 0; & 2a_2 N_2(t)(b_0 p_0(t) - b_1 p_1(t)) > 1 \\ 1; & 2a_2 N_2(t)(b_0 p_0(t) - b_1 p_1(t)) < 1 \end{cases}
\tag{36}
$$

$$
y(t) = \begin{cases} 1; & p_1(t) > p_0(t) \\ y_m; & p_1(t) < p_0(t) \end{cases}
\tag{37}
$$

where the costate vector satisfies the following set of equations,

$$
\begin{aligned}
\dot{p}_0(t) &= y(t) a_0 (p_0(t) - p_1(t)), & p_0(T) &= r_0 , \\
\dot{p}_1(t) &= a_1 (p_1(t) - p_2(t)), & p_1(T) &= r_1 , \\
\dot{p}_2(t) &= a_2 [p_2(t) - 2u(t)(b_0 p_0(t) + b_1 p_1(t))] , & p_2(T) &= r_2 ,
\end{aligned}
\tag{38}
$$

The arising two-point boundary value problem (TPBVP) expressed by equations (34) and (36–38) is formally similar to the TPBVP expressed by equations (26) and (28–30) and leads to the same mathematical problems. In this case the analysis of singular arcs is slightly more cumbersome [125]. For constant y we have:

$$
\frac{\partial}{\partial u} \frac{d^2}{dt^2} \frac{\partial H}{\partial u} = 4a_1 a_2 b_1 > 0
\tag{39}
$$

violating the Legendre-Clebsch condition. These calculations therefore exclude the optimality of singular controls u when y is constant. It might still be possible, however, that y is singular and not constant over any subinterval $J \subset I$. In this case u also must be singular on I. For this example the Goh condition is actually satisfied but after some simple but lenghty calculations we have found that it is possible only for $u = 0.5$ and leads to constant N_is and p_is and in consequence to constant y but it, in turn, implies violation of the Legendre-Clebsch condition. In [125] it is shown also that the sufficient conditions for bang-bang strategies in this case and in more general class of multicompartment models could be derived similarly as for the previously considered cases. In this case the numerical results can be obtained by the same gradient method [124].

The interesting finding [128] is that our results do not change at least in qualitative sense if instead of modeling and minimizing cancer population we rather decide to model and maximize population of cells in critical normal tissues while maximizing the cumulative negative cytotoxic effect.

4 Evolution of Resistance by Gene Amplification

4.1 Biological Background

The amount of DNA per cell remains constant from one generation to another because during each cell cycle the entire content of DNA is duplicated and then at each mitotic cell division the DNA is evenly apportioned to two daughter cells. However, recent experimental evidence shows that for a fraction of DNA, its amount per cell and its structure undergo continuous change.

One way the genome of cancer cells may rapidly evolve is by an increase in copy number of specific genes, referred to as *gene amplification*. Gene amplification can be enhanced by conditions that interfere with DNA synthesis and is increased in some mutant and tumor cells. Increased number of gene copies may produce an increased amount of gene products and, in tumor cells, confer resistance to chemotherapeutic drugs. Amplification of oncogenes has been observed in many human tumor cells and also may confer a growth advantage on cells which overproduce the oncogene products (for an overview see surveys by Stark [109] and Windle and Wahl [146]).

In the classical experiments of Schimke and his coworkers [23, 61], the anticancer drugs served to select for cells with amplified genes. In some of cell lines, when the selective agent was removed, the cells with amplified genes gradually disappeared from the population. The stochastic mechanism leading to this reversal is discussed in more detail further in this subsection. It was· observed that in such cases the amplified genes were located on extrachromosomal fragments of DNA called *Double Minute Chromosomes (DM's)*. In other cases, the amplification was stable, ie. persisted after the selective agent had been removed. In such cases, the amplified genes usually are located on elongated chromosome arms. The most regular of these elongated arms exhibit a regular band structure (the so called *Homogeneously Staining Regions* or *HSR's*), but other less regular structures are also observed. They are either caused by reintegration of extrachromosomal genes as proposed by Windle, Wahl and coworkers [145], or they arise by a separate mechanism as proposed by Stark and coworkers [107]. Mathematical models show that depending on circumstances each of the two variants of stable amplification is plausible [7, 70] (see also a critique by Harnevo and Agur [54]).

4.2 Probabilistic Modeling of Unstable and Stable Gene Amplification

Unstable Gene Amplification

Summary of Observations. In some populations of cells with double minute chromosomes, both the increased drug resistance and the increase in number of gene copies are *reversible*. The classical experiment confirming this includes transfering the resistant cell line into drug-free medium, [23, 61], where cells gradually lose resistance to the drug by losing extra gene copies. In these experiments, the dihydrofolate reductase (DHFR) gene was amplified after exposing murine 3T6 cells [23] or mouse sarcoma S-180 cells [61] to Methotrexate (MTX).

The population distribution of numbers of gene copies per cell can be estimated by flow cytometry after staining gene products. In the experiments mentioned, [61], two features of these distributions are notable. (1) As expected, the proportions of resistant cells (with amplified genes) decrease with time. (2) Less obvious, the shape of the distribution of the number of gene copies limited to the resistant cell subpopulation seems to remain stable during the loss of resistance.

The Branching Random Walk and Other Models. A mathematical model of unstable drug resistance should take into account (1) stochastic changes in number of gene copies from one generation to another and (2) the stochastic variability in cell lifetimes. One stochastic process which accomodates both (1) and (2) is a random walk superimposed on the time-continuous branching process [6] of cell proliferation, ie. a *branching random walk* [71]. We consider a population of abstract particles of types $j = 0, 1, 2, \ldots$:

1. The lifespans of all particles are independent identically distributed exponential random variables with mean $1/\lambda$.
2. At the moment of death, a particle of type $j \geq 1$ produces two progeny particles each belonging to type $j + 1$ with probability b, to type $j - 1$ with probability d, and to type j with probability $1 - b - d$. A particle of type $j = 0$ produces two progeny of type 0.
3. The process is initiated at time $t = 0$ by a single particle of given type i.

The simplest models of gene amplification in [68] and [71] assume the above process. Cells with 2^{j-1} gene copies are said to belong to type j (with 0 gene copies, to type 0). The parameters b and d are the probabilities of gene *amplification* and *deamplification*, respectively.

One of the properties of Markov processes with absorbing states is the possibility of existence of the quasi-stationary distributions. In intuitive terms, the unabsorbed part of the probability mass of the process, while constantly shrinking, approaches a limit if it is properly normed. The Yaglom theorem for subcritical branching processes [6] can be quoted as an example. It is this property that explains the apparent stability of distributions of gene copy number per cell in the resistant subpopulation, placed in the non-selective medium.

Indeed, let us assume the time-discrete equivalent of the model outlined above (as in [68]). Let us denote by X_n the *type* (see the definition above) of the cell in the nth generation of a randomly selected lineage ($n \geq 0$). Then, $\{X_n, \ n \geq 0\}$ is a time-discrete Markov chain with the following transition matrix

$$
\begin{bmatrix}
1 & 0 & 0 & 0 & 0 & \cdots \\
d & (1-b-d) & b & 0 & 0 & \cdots \\
0 & d & (1-b-d) & b & 0 & \cdots \\
0 & 0 & d & (1-b-d) & b & \cdots \\
 & & & \ddots & \ddots & \cdots \\
0 & 0 & 0 & & & \\
\cdots & \cdots & \cdots & \cdots & \cdots & \cdots
\end{bmatrix} . \tag{40}
$$

$\{X_n, \ n \geq 0\}$ is a random walk with an absorbing boundary at 0. Let us denote by p_i^n the probability that the cell type is i in the nth cell generation,

$$
p_i^n = \Pr\{X_n = i\} . \tag{41}
$$

We consider the limit properties, as n tends to infinity, of the gene extinction probability p_0^n, and of the set of conditional probabilities,

$$
c_i^n = \Pr\{X_n = i | X_n \neq 0\} = p_i^n/(1 - p_0^n) . \tag{42}
$$

that the cell type is i, provided the gene is not extinct. We limit ourselves to the most important subcritical case, $d > b$, and assume $X_0 = 1$. Let $v_n(s) = \sum_{i \geq 1} c_i^n s^i$ be the conditional probability generating function (*p.g.f.*) of X_n, given $X_n \neq 0$ and let $E_\infty = \lim_{n \to \infty} \sum_{i \geq 1} 2^{i-1} c_i^n$ be the expected number of gene copies as $n \to \infty$. We have the following result (Theorem 1 in [68]).

Suppose that $d > b$. Then

$$
(1 - p_0^n) \sim Kh^n/\sqrt{n^3} , \tag{43}
$$

as $n \to \infty$, where $h = 1 - (\sqrt{d} - \sqrt{b})^2$, $K = [1 - (\sqrt{d} - \sqrt{b})^2]^{3/2} \sqrt[4]{d} \ /[2\sqrt{\pi}(\sqrt{d} - \sqrt{b})^2 \sqrt[4]{b^3}]$, and

$$v_n(s) \to v(s) \equiv s \left(\frac{\sqrt{d} - \sqrt{b}}{\sqrt{d} - s\sqrt{b}} \right)^2 , \tag{44}$$

and consequently,

$$E_\infty = v(2)/2 = \left(\frac{\sqrt{d} - \sqrt{b}}{\sqrt{d} - 2\sqrt{b}} \right)^2 , \tag{45}$$

as $n \to \infty$.

This result can be derived from Theorem 2 and Lemma 3 in [95]. As stated above, it assures existence of the limit distribution of the number of gene copies per cell in selective conditions.

Stable Amplification

Summary of Observations. In the experimental system of Windle, Wahl and co-workers [145], the amplification of the DHFR gene was observed in a Chinese Hamster Ovary (CHO) cell line, which contained only a single DHFR gene. Cells were challenged by MTX. Amplified genes residing on extrachromosomal elements were observed in cell cultures 8-9 generations later, while predominantly chromosomally amplified genes were seen after about 30 generations (only these two time points were investigated). This can be interpreted as an indication that some extrachromosomal elements containing amplified gene copy numbers are eventually reintegrated into chromosomes.

Mathematical Model and Its Predictions. In the model devised to reproduce these observations [70], the basic indivisible unit which serves as the template for the production of additional gene copies is *the amplicon*, which contains at least one copy of the target gene. The size of such structures could range from submicroscopic to an entire arm of a chromosome and they may be circular or linear. The *acentric (replicating) element (ARE)* is understood to be an extrachromosomal molecular structure containing one or more amplicons but no centromere. A centromere is required for regular segregation to daughter cells. *The reintegrated element (RE)* is the ARE after it has reintegrated into a chromosome.

The following processes are considered in the model: (a) change in the number of ARE's per cell, (b) change in the number of amplicons per ARE, and (c) reintegration of ARE's into chromosomes.

Types of elements: ARE's containing $i = 1, 2, \ldots$ amplicons, and RE's containing $i = 1, 2, \ldots$ amplicons. In each cell generation, with probability a, the ARE containing i amplicons replicates to yield a product with $2i$ amplicon copies. The catenated replication product then dissociates producing two acentric molecules. This process results in a pair of molecules containing, respectively, j and $2i - j$ amplicons, where $j = 1, \ldots, 2i - 1$. It is assumed that the probability of each pair $(j, 2i - j)$ is the same, equal to $1/(2i - 1)$.

The molecules segregate so that they both go to the same daughter cell with probability q, and go to different daughter cells with probability $1 - q$. With probability g, the ARE with i amplicon copies replicates to yield a product with $2i$ amplicon copies, but this replication product does not dissociate. It then goes with equal probability to one of the two daughters. With probability $c = 1 - (a + g)$, per cell generation, the ARE containing i copies of the amplicon, integrates into a chromosome with a centromere and then replicates and segregates with the chromosome. This results in each daughter cell containing an equal number of RE copies. The probability of reintegration is $c = 1 - (a + g)$.

We may formally define the following random variables:

- $X_n^i(\omega)$, the number of ARE's with i copies of the amplicon, in the n-th cell generation,
- $Y_n^i(\omega)$, the number of RE's with i copies of the amplicon, in the n-th cell generation.

The sequence $\{\{(X_n^1, Y_n^1), (X_n^2, Y_n^2), \ldots\}, n = 0, 1, 2, \ldots\}$, is a *multitype Galton-Watson process with a denumerable infinity of particle types*. [66].

Modeling the expected values of the process enables reproducing the main features of Wahl's experiments: (1) The initial increase in number of acentric elements per cell, and the number of amplicon copies per acentric element. (2) Subsequent decrease of the number of ARE's per cell, as they become reintegrated. (3) Eventual emergence of a population of cells containing only integrated elements with a spectrum of amplicon copy numbers at one or more chromosomal locations.

5 Control Under Evolving Resistance

5.1 Clonal Resistance/Simulation of Gene Amplification

Resistance to antineoplastic drugs has been a major impediment to the successful treatment of cancer. Recent studies suggest that several mechanisms are responsible for the emergence of drug resistance and that high levels of resistance and poor prognosis are strongly associated with gene or oncogene amplification.

In recent years the problem of drug resistance in cancer has been mathematically attacked by many authors. The first series of models were devised by Coldman and Goldie [30] (for an overview see the book by Wheldon [142]). Underlying these models was the assumption that drug resistance in cancer results from a single mutational event whose probability is constant and independent of external constraints. The model was generalized to describe evolution of resistance to a number of agents. If it is assumed that multiple resistance is the most important thing to avoid in the course of chemotherapy, then the resulting recommendation is to alternate treatments effective against strains resistant to single agents, as frequently as possible [46].

Harnevo and Agur [53] introduce a model which treats the emergence of drug resistance as a dynamic process rather than a single event. Using this model, based on their previous works [52], they focus on gene amplification as one of the mechanisms that may lead to drug resistance, and show how changes in the underlying assumptions affect the predictions about treatment efficacy. The mathematical modeling results suggest that under gene amplification dynamics with high amplification probability, protocols involving frequent low-concentration dosing may result in the rapid evolution of large fully resistant residual tumors; the same total doses divided into high-concentration doses applied at larger intervals may result in partial or complete remission. This last recommendation is an alternative to that of Coldman and Goldie.

Another suggestion is that treatment prognosis may be largely improved if cells bearing a large number of gene copy number have high mortality. Therefore, it may be interesting to examine the possibility of incorporating in the treatment an agent (hypothetical, at present) that increases the mortality of cells carrying highly amplified genomes.

5.2 Mathematical Model and Optimization of Control Under Evolving Resistance

In this subsection we present an infinite system of differential equations which may be used to model controling a cell population with evolving drug resistance caused by gene amplification or other mechanisms. The model is general enough to accomodate different interpretations (see further on). The model is motivated by a representation in the terms of the branching random walk [71], but it also can be understood as a mathematical variation of the model used by Harnevo and Agur in [53].

The hypotheses are as follows: We consider a population of cells of types $i = 0, 1, 2, \ldots$ Cells of type 0 are sensitive to the agent, whereas the types $i = 1, 2, \ldots$ consist of resistant cells of increasing level of resistance (for example, with increased number of DHFR or CAD gene copies per cell).

1. The lifespans of all cells are independent identically distributed exponential random variables with means $1/\lambda_i$ for cells of type i.
2. A cell of type $i \geq 1$ may mutate in a short time interval $(t, t + dt)$ into a type $i + 1$ cell with probability $b_i dt + o(dt)$ and into type $i - 1$ cell with probability $d_i dt + o(dt)$. A cell of type $i = 0$ may mutate in a short time interval $(t, t + dt)$ into a type 1 cell with probability $\alpha dt + o(dt)$, where α is several orders of magnitude smaller than any of b_is or d_is.
3. The chemotherapeutic agent affects cells of different types differently. It is assumed that its action results in fraction u_i of ineffective divisions in cells of type i.

4. The process is initiated at time $t = 0$ by a population of cells of different types.

The postulated relationship for the rate α of the primary amplification event can be written as follows

$$\alpha \ll \min(d_i, b_i), \quad i \geq 1 . \tag{46}$$

In view of the subcriticality of the process in Subsect. 4.2, it seems reasonable to assume

$$d_i > b_i, \quad i \geq 1 . \tag{47}$$

If we denote $N_i(t)$ the expected number of cells of type i at time t, we obtain the following infinite system of differential equations:

$$\begin{cases} \dot{N}_0(t) = [1 - 2u_0(t)]\lambda_0 N_0(t) - \alpha N_0(t) + d_1 N_1(t) , \\ \dot{N}_1(t) = [1 - 2u_1(t)]\lambda_1 N_1(t) - (b_1 + d_1)N_1(t) + d_2 N_2(t) + \alpha N_0(t) , \\ \cdots \\ \dot{N}_i(t) = [1 - 2u_i(t)]\lambda_i N_i(t) - (b_i + d_i)N_i(t) + d_{i+1}N_{i+1}(t) \\ \qquad + b_{i-1}N_{i-1}(t), \quad i \geq 2 , \\ \cdots \end{cases} \tag{48}$$

Also, the folowing relationships between b_is and d_is seem to be justified by the intuition that cells overloaded with amplified gene copies may aquire new copies with more difficulty and lose them easier:

$$d_{i+1} \geq d_i, \quad b_{i+1} \leq b_i, \quad i \geq 1 . \tag{49}$$

As postulated by Schimke (see e.g. [23, 61]), cells with more copies of the drug resistance gene may proliferate slower, ie.

$$\lambda_{i+1} \leq \lambda_i, \quad i \geq 0 . \tag{50}$$

In the simplest case, in which the resistant cells are totally insensitive to drug's action, and if we ignore differences between parameters of cells of different type, i.e.

$$u_0 = u, u_i = 0, i \geq 1, \quad \text{and} \quad b_i = b, d_i = d, \lambda_i = \lambda, i \geq 0 ,$$

the system (48) assumes the following form:

$$\begin{cases} \dot{N}_0(t) = [1 - 2u(t)]\lambda N_0(t) - \alpha N_0(t) + dN_1(t) , \\ \dot{N}_1(t) = \lambda N_1(t) - (b + d)N_1(t) + dN_2(t) + \alpha N_0(t) , \\ \cdots \\ \dot{N}_i(t) = \lambda N_i(t) - (b + d)N_i(t) + dN_{i+1}(t) + bN_{i-1}(t), i \geq 2 , \\ \cdots \end{cases} \tag{51}$$

Note that in this model d and b denote respective intesivities and not probabilities as it was in matrix (40). Model (51) may be used to find the optimal control which minimizes an appropriate performance index, e.g.

$$J = \sum_{i \geq 0} r_i N_i(T) + \int_0^T u(t)dt , \qquad (52)$$

Necessary conditions for optimal control could be found using the maximum principle in its abstract version (see e.g [91]). They are formally similar to those obtained for respective finite dimensional problems. To solve them efficiently, finite approximation of the system should be used. The other possibility conferred for example in [131] is to reconfigure the model (51) into the equivalent integro-differential form.

The model (48) can describe dynamics of any cell population stratified into a sequence of subcompartments (types) with different kinetics, with fluxes of cells between these subcompartments. For example, the amplified gene may not confer resistance but allow the cell to proliferate faster (or to alter the pattern in a more complex way, see e.g. [67]). This might be the case if it is an oncogene. Then we might assume all u_is equal and $\lambda_{i+1} \geq \lambda_i$.

Systems of the type (48) and (51) are not as straightforward as finite dimensional systems of differential equations. However, at least in simpler cases, their asymptotic behavior can be characterized quite precisely. As an example, let us consider the following system,

$$\begin{cases} \dot{N}_1(t) = \lambda N_1(t) - (b+d)N_1(t) + dN_2(t), \\ \qquad \cdots \\ \dot{N}_i(t) = \lambda N_i(t) - (b+d)N_i(t) + dN_{i+1}(t) + bN_{i-1}(t), \; i \geq 2 , \\ \qquad \cdots \end{cases} \qquad (53)$$

This is a model of population of cells in which the sensitive cells are instantly annihilated, and there is no influx of new resistant cells. Let us denote $N(t) = \sum_{i \geq 1} N_i(t)$. We have the following result obtained using the methods of [71]. Suppose that $N_i(0) = \delta_{i1}$ and $d \neq b$. Then

$$N(t) = e^{\lambda t} - e^{\lambda t} \sqrt{d/b} \int_0^t \frac{I_1(2\sqrt{bd\tau})}{\tau} e^{-(b+d)\tau} d\tau . \qquad (54)$$

where $I_1(t)$ is the modified Bessel function of order 1 [1]. Moreover,

$$N(t) \sim \left[1 - \frac{\min(b,d)}{b} \right] e^{\lambda t}$$

$$+ \frac{d}{2\sqrt{\pi} \sqrt[4]{(bd)^3}(\sqrt{d}-\sqrt{b})^2} t^{-3/2} e^{[\lambda-(\sqrt{d}-\sqrt{b})^2]t} , \qquad (55)$$

as $t \to \infty$.

This result may be derived using Laplace transform machinery. Denote Laplace transforms of $N_1(t)$ and $N(t)$ by $\hat{N}_1(s)$ and $\hat{N}(s)$, i.e. $\hat{N}_1(s) = \int_0^\infty N_1(t)e^{-st}dt$, $\hat{N}(s) = \int_0^\infty N(t)e^{-st}dt$. Then:

$$\hat{N}_1(s) = \frac{s - \lambda + b + d - \sqrt{(s - \lambda + b + d)^2 - 4bd}}{2bd}. \tag{56}$$

$$\hat{N}(s) = -\frac{s - \lambda + b + d - \sqrt{(s - \lambda + b + d)^2 - 4bd}}{2b(s - \lambda)} + \frac{1}{s - \lambda}. \tag{57}$$

Note that (see [36]).

$$(s + b + d) - \sqrt{(s + b + d)^2 - 4bd}$$

is the Laplace transform of

$$(2\sqrt{bd}/t)I_1(2\sqrt{bd}t)\exp[(-b - d)t],$$

Using this we obtain time functions $N_1(t)$ (see below) and $N(t)$ (see(54)) by performing inverse Laplace transforms of (56) and (57):

$$N_1(t) = \frac{I_1(2\sqrt{bd}t)}{t\sqrt{bd}}e^{[\lambda - (b+d)]t}, \tag{58}$$

Using the equations (58) and (54) we can analyze the behavior of functions $N_1(t)$ and $N(t)$ as t approaches infinity. The formulae for the asymptotic expansions of $I_1(t)$ and $\int_0^t \frac{I_1(2\sqrt{bd}\tau)}{\tau}e^{-(b+d)\tau}d\tau$ given in Lemma 1 and Lemma 2 in the paper [71] (obtained via the Laplace method for integrals [24]) lead to the asymptotic expansions for $N_1(t)$:

$$N_1(t) \sim \frac{1}{2\sqrt{\pi}\sqrt[4]{(bd)^3}}t^{-3/2}e^{[\lambda - (\sqrt{d} - \sqrt{b})^2]t}, \tag{59}$$

and $N(t)$ (see(55)). From (59) and (55), the condition that both $N_1(t)$ and $N(t)$ converge exponentially to zero, as $t \to \infty$, is:

$$\sqrt{d} - \sqrt{b} > \sqrt{\lambda}. \tag{60}$$

If (60) is not satisfied, then we have two possibilities. If $\sqrt{d} - \sqrt{b} < \sqrt{\lambda}$, then the solution diverges exponentially to infinity. If $\sqrt{d} - \sqrt{b} = \sqrt{\lambda}$, then both N_1 and N still converge to zero. However, the convergence is not exponential [73]. Let us notice that the term at $e^{\lambda t}$ in the asymptotic expansion disappears if $d > b$. This separates the behavior in the supercritical case from that in the subcritical case. In the former, the resistant population grows exponentially. In the latter, it decays only if $\sqrt{d} - \sqrt{b} > \sqrt{\lambda}$. The same reasoning could be repeated for initial conditions of the form $N_i(0) = \delta_{ik}$ for $k > 1$ leading to similar expression for $N(t)$ and similar asymptotic behaviors (see [129]). Since the system is linear the condition (60) is necessary and sufficient for eradication of the resistant subpopulation for any initial conditions with final support. If λ is considered the only parameter affected by control, this means that unless somehow accessed by cytostatics, the resistant subpopulation may maintain itself even in the subcritical case. The analysis presented in [99] leads to a conclusion that being stable for any finite initial condition, the solution to (53) can diverge if we allow initial conditions with infinitely many nonzero elements. The factor that determines stability of the solution in this case is that the rate of decay of successive elements of the initial vector. The rate of decrease must be faster than $[b + d - \lambda - \sqrt{(b + d - \lambda)^2 - 4bd}]^{-1}$ [73]. Biologically, the results for initial conditions with infinite support can be interpreted as follows: Suppose that a significant subpopulation of resistant cells reached large number of gene copies. Then this population is a persistent source of proliferating malignant cells, much more difficult to eradicate than it would be the case under a finite mutation model [129].

Asymptotic analysis of this model results in some understanding of its dynamics . This, in our opinion, is the first step towards a more rigorous mathematical treatment of the dynamics of drug resistance and/or metastasis In this case the system is decomposed onto two parts one which includes only the sensitive subpopulation modeled by the first equation of (51) and the remaining infinite dimensional part of this model which describes the drug resistant subpopulation. The results on the asymptotic behavior of the drug resistant subpopulation not supplied from the sensitive compartment, combined with the Laplace transforms machinery and theory of closed loop systems with positive feedback (Nyquist theorem) enable analysis of an asymptotic behaviour of the whole population modeled by (51) for the case of constant dosage drug administration. More precisely using Nyquist theorem [147] for systems with irrational transfer functions we found [99] that the conditions which ensure asymptotic eradication of the overall cancer population by continuous constant drug dosage in subcritical case are given by (60), and

$$u > \frac{\alpha}{d(-b + d - \lambda + \sqrt{(b + d - \lambda)^2 - 4bd})} + \frac{1}{2} . \tag{61}$$

Moreover the representation of the cancer population in the form of the closed loop system with positive feedback enables reformulution of the infinite dimensional model of dynamics into the integro-differential form [130]. This in turn leads to formulation of optimization problem for protocols design in the presence of drug resistance [131] in the form treatable by an abstract Pontryagin maximum principle in the version given by [10]. In the simplest case when the therapy is initiated when there are no resistant clones yet and weights $r_i = r$ for $i = 0$ and r_1 elsewhere we have the following reformulation of the model (51) and the performance index (52):

$$\begin{aligned} N_0(t) &= d\alpha \int_0^t \phi(t - s)N_0(s)ds - \alpha N_0(t) + (1 - 2u(t))\lambda N_0(t), \quad t > 0 \\ N_0(t) &= N_0(0), \quad\quad\quad\quad\quad\quad\quad\quad\quad\quad\quad\quad\quad\quad\quad\quad\quad\quad\quad\quad t \leq 0 \end{aligned} \tag{62}$$

$$J = rN_0(T) + \int_0^T [r_1 N(T - t)N_0(t) + u(t)]dt \,, \tag{63}$$

where $\phi(t)$ is given by formula (58) for $N_1(t)$ and $N(t)$ by formula (54). Using the abstract maximum principle we obtain the following necessary conditions for optimal control $u(t)$ and costate $p(t)$ [131]:

$$u(t) = \begin{cases} 0; & 2\lambda N_0(t)p(t) < 1 \\ 1; & 2\lambda N_0(t)p(t) > 1 \end{cases} \tag{64}$$

$$\dot{p} = -\left[d\alpha \int_t^T \phi(s - t)p(s)ds + p(t)((1 - 2u)\lambda - \alpha) + \alpha N(T - t) \right], \quad p(T) = r \tag{65}$$

This analysis does not take into account singular solutions for which in this case we have no results dealing with their elimination. The gradient method proposed for two and three compartment models [37, 38] may be however used with some technical modifications [105] to find optimal bang-bang and suboptimal periodic solutions. The models in this section are based on the hypothesis that the process of gene amplification can be described by a branching random walk, as in [71]. A more realistic process, including proliferation of gene copies between cell divisions, as well as random segregation of gene copies between daughter cells, is described in [66].

6 Remarks on Estimation of Parameters

6.1 Estimation of Cell Cycle Parameters and Drug Action in Cultured Cells

Much work has been done on estimation of cell cycle transit times and of the fractions of cells arrested and/or killed by drugs in cultured normal and transformed cells. The most systematic series of experiments and measurements known to us has been carried out in the 1980's in the laboratory of Darzynkiewicz, Traganos and their coworkers in the Memorial Sloan-Kettering Cancer Center and in the New York Medical College. A number of cultured cell lines and a variety of drugs in various concentrations were evaluated by these researchers using flow cytometric techniques.

The following account is based on reviews [33] and [138] which include numerous original references. The stathmokinetic or "metaphase arrest" technique consists of blocking cell division by an external agent (usually a drug, e.g. vincristine or colchicine). The cells gradually accumulate in mitosis, emptying the postmitotic phase G_1 and with time also the S phase. Flow cytometry allows precise measurements of the fractions of cells residing in different cell cycle phase. The pattern of cell accumulation in mitosis (M) depends on the kinetic parameters of the cell cycle and is used for estimation of these parameters. Exit dynamics from G_1 and transit dynamics through S and G_2 and their subcompartments can be used to characterize very precisely both unperturbed and perturbed cell cycle parameters. A true arsenal of methods have been developed to analyze the stathmokinetic data. Application of these methods allow quantification of the cell-cycle-phase action of many agents.

One of the interesting findings was the existence of *aftereffects* in the action of many cytotoxic agents. The action of these drugs may extend beyond the span of a single cell cycle. For example, cells blocked in the S-phase of the cell cycle and then released from the block, may proceed apparently normally towards mitosis but then fail to divide, or divide but not be able to complete the subsequent round of DNA replication. In some experiments it was possible to trace the fates of individual cells and conclude that their nuclear material divided but the cytoplasmic contents failed to separate.

The consequence of the aftereffects is that it may be difficult to infer the long-term effects of cytotoxic drugs based on short term experiments like the stathmokinetic experiment. One way of testing this assertion is to carry out both types of experiments, short term and long term, subjecting cells to the action of the same concentration of the same drug. We may then estimate the parameters of the cell cycle and of drug action based on the short-term experiment, substitute them into a mathematical model and try to predict the results of the long-term experiment. This program has been carried out in a study by Kimmel and Traganos [74]. Using two concentrations of an experimental anticancer agent CI-921, it was found that while cell cycle estimates based on stathmokinetic experiments did not differ for these two concentrations, the effects of continuous 24-hour exposure to the drug were completely different. Only the effects of low concentration continuous exposure were predicted by the mathematical model using estimates from stathmokinesis.

Analogous aftereffects following irradiation of cells were discovered by Kooi and co-workers [76].

Another long term program of estimation based on flow cytometry has been carried out by Bertuzzi and Gandolfi and their collaborators at IASI in Rome and European Institute of Oncology. For many years their interest has been concentrated mainly on reducing errors of estimates using procedures like regularization, but also simulations and modeling of cytotoxic action on the cell cycle (see e.g. [18, 20]). Moreover their estimation procedures is also used for modelling the cell kinetic characteristics of in vivo experimental tumors [18]. Recently they have also been concentrated on the modeling of tumor cords (see e.g. [17, 19]). Estimation of parameters for such models results in better understanding of nongenetic reasons of drug resistance.

6.2 Estimation of Cell Cycle Parameters in Cells from Human Tumors

Recently, much research has been carried out on estimation of cell kinetic parameters of cells in human tumors in vivo. Basically, the procedure consists of injecting the tumor with a labeling compound selectively incorporated by cells synthesizing DNA and then, after removal of the tumor, of following the "relative movement" of the labeled cells through the S-phase. The method was introduced by Begg and co-workers [13]. They followed-up with a series of application papers [14, 15, 55]. Their main interest is on the potential of pre-treatment cell kinetic parameters to predict outcome in cancer patients treated by radiotherapy. One of the findings is that pretreatment cell kinetic measurements carried out using flow cytometry only provide a relatively weak predictor of outcome after radiotherapy e.g. [11, 12, 56]. One of possible reasons of this negative result is the change of the parameters during radiotherapy and the effect of breaks in the therapy for the cell cycle parameters. In our collaborative research with colleagues from MCS Institute of Oncology in Gliwice we have observed such phenomena while analysing similar material (neck and head cancers) [135, 136]. It is difficult to extend these results for chemotherapeutic effects.

A series of mathematical refinements and applications of the Begg's method have been introduced by R.A. White of the Biomathematics Department at the M.D. Anderson Institute in Houston [143, 144]. Of a number of papers by other authors, we quote one devoted to cell cycle kinetics of leukemias [101]. There are also available many papers on parameter estimation of cell cycle kinetics for experimental human tumors transplanted in mice or other animals (see e.g. previosuly mentioned [18], where results for human ovarian carcinoma transplanted in mice are presented).

New possibilities in cell cycle parameter estimation both in vitro and in vivo are now established by DNA microarray technology. By processing the data on expression of thousands of genes in different time samples one can identify the dynamics behavior of the analysed cell populations. There have been a number of bioinformatical and biomathematical tools developed to cope with such analysis. Among them we refer to the one based on Singular Value Decomposition which seems to give especially promising results (see, [104] and references therein).

6.3 Estimation of Rates of Emergence and Evolution of Resistance

Based on gene amplification studies, there exist three phases in the evolution of resistance process:

- The relatively rare *primary event*, i.e. the establishment of the founder cell of the resistant clone containing at least one unstable copy of the target gene (the probability of this event, per cell division, corresponds to the ratio α/λ in our (48)).
- Subsequent *amplification* and *deamplification* events, occuring at high rates compared to α/λ, resulting from instability of the amplified gene (the probabilities of these events, per cell division, correspond to the ratios b_i/λ and d_i/λ in (48)).

- Possible *stabilization* of the resistant phenotype, by integration of the amplified gene in the chromosomal structures (no counterpart in (48)).

The numerical values of the probabilities of gene amplification and deamplification can be estimated based on data in [23] and [61]. The probabilities of deamplification (d) are of the order of 0.10 in both cases, while the probabilities of amplification (b) are about 5 times lower. The process is strongly subcritical. This means among others that in the absence of selection, the amplified phenotype disappears from the population. It can be revived by rare *primary events*, such as amplification of extrachromosomal genes following a deletion of the target gene from the chromosome arm (see further on). The primary tool with which the primary rate was estimated, is the Luria-Delbrück's fluctuation analysis [82]. It consists of finding an experimental distribution of the number of mutant (i.e. resistant) colonies in cell populations cultured for a number of generations, and fitting it to the theoretical distribution derived based on a branching process-type model of proliferation and mutation. A number of researchers carried out this procedure for mutation to drug resistance (by gene amplification and other means) [88, 89, 137, 139], obtaining estimates of the mutation probabilities, per cell division, in the range from 10^{-8} to 10^{-6}, with generally higher estimates for tumorigenic than for "normal" cells. The data from the above papers were re-analyzed in a paper by Kimmel and Axelrod [69], using a two-stage model of mutation. Although the estimates of primary event probabilities remain mostly unchanged, the probabilities of second stage forward and backward mutation are much higher, comparable to the estimates of amplification and deamplification probabilities obtained in [68] and in [71] (of the order of 0.02 and 0.10, respectively, as mentioned above).

The classical explanation for the loss of resistance in cells with amplified DNA in extrachromosomal elements is that in the absence of selective pressure cells with extra gene copies grow slower and are outgrown by the sensitive cells [61]. Our model assumes a purely stochastic mechanism. A combination of two mechanisms is likely. For further comments, see [68].

Estimates of the reintegration probabilities are provided in the study by Kimmel, Axelrod and Wahl [70] where fitting the model to data from [145] makes possible to estimate its parameters a, g, c, and q. The best fitting values are $a = 0.780$, $g = 0.195$, $c = 0.025$, and $q = 0.9$ The rate of integration c and the probability of cosegregation q are biologically important parameters. A high rate of cosegregation might have clinical implications in the sense of making some cells more sensitive to chemotherapy.

7 Discussion

In this chapter we describe the cell-cycle-phase dependence of cytotoxic drug action and drug resistance in the context of optimization of cancer chemotherapy.

Attempts at optimization of cancer chemotherapy using optimal control theory have a long history ([113] is a review by Swan). The methods include the maximum principle both in the discrete [8, 64] (with parameters taken from [51, 63]) and continuous versions [103, 116], for a variety of models and performance indices. The idea has been criticized many times (see e.g. [134, 142]). Only simplest concepts have won attention in the medical world. These include the clonal resistance model [46] and the kinetic resistance theory by Norton and Simon [93].

The simplest cell-cycle-phase dependent models of chemotherapy can be classified based on the number of compartments and types of drug action modeled. In all these models the attempts at finding optimal controls are confounded by the presence of singular and periodic trajectories, and multiple solutions but recently singular trajectories are excluded and sufficient conditions for strong local optimality are found for a class of bang-bang strategies. Morover, efficient numerical methods have been developed. In simpler cases, it is possible to provide exhaustive classification of solutions. We have reviewed analytic and computational methods which are available. All these attempts have to be viewed with caution, because of the existence of *aftereffects* in the action of many cytotoxic agents. The action of these drugs may extend beyond the span of a single cell cycle. For example, cells blocked in the S-phase of the cell cycle and then released from the block, may proceed apparently normally towards mitosis but then fail to divide, or divide but not be able to complete the subsequent round of DNA replication. If such effects are substantial, they are likely to disrupt or complicate the resonances. As indicated for example in [96, 97], the aftereffects due to accumulation of drugs (in this case MTX) result in great interindividual differences of the effectiveness of treatment.

The consequence of the aftereffects is that it may be difficult to infer the long-term effects of cytotoxic drugs based on short term experiments like the stathmokinetic experiment. One way of testing this assertion is to carry out both types of experiments, short term and long term, subjecting cells to the action of the same concentration of the same drug. We may then estimate the parameters of the cell cycle and of drug action based on the short-term experiment, substitute them into a mathematical model and try to predict the results of the long-term experiment. Constructing mathematical models including aftereffects is possible but leads to notational and computational complications. Essentially, part of cells released from the direct action of the drug, are redirected into a different cell cycle in which a large part of them are either permanently arrested or die. This leads to models with increased dimensionality. It seems, however, that it still is possible to place the models in the general class (P) of multicompartmental models discussed in [125].

Concerning the emergence of drug resistance, we have presented the problem in the framework of gene amplification, although much of what is written may apply to different mechanisms which are reversible and occur at high frequency. We have defined a mathematical model which can be used to pose and solve an optimal chemotherapy problem under evolving resistance. We have shown some results regarding dynamics of this model and techniques used to find solutions for optimization of chemotherapy protocols. Analysis of variants of this model should give insight into possible scheduling strategies of chemotherapy in the situations when drug resistance is a significant factor. It is possible, for example, using the decomposition technique presented in the paper to include both effects of drug resistance and phase specificity or partial resistance of different subpopulations both in singleagent or multiagent chemotherapy (see [106]).

All possible applications of the mathematical models of chemotherapy are contingent on our ability to estimate their parameters. There has been a progress in that direction, particularly concerning precise estimation of drug action in culture and estimation of cell cycle parameters of tumor cells in vivo. Also, more is known about the mutation rates of evolving resistant cell clones.

The traditional area of application of ideas of cell synchronization, recruitment and rational scheduling of chemotherapy including multidrug protocols, is in treatment of leukemias. It is there where the cell-cycle-phase dependent optimization is potentially useful.

The emergence of resistant clones is a universal problem of chemotherapy. However, it seems that its most acute manifestation is the failure to treat metastasis. A part of this problem is the imperfect effectiveness of adjuvant chemotherapy as the tool to eradicate undetectable micrometastases. In view of toxicity of anticancer drugs, optimal scheduling is potentially useful in improving these treatments. Yet another challenge discussed recently in modeling of cancer chemotherapy is related to antiangiogenic therapy (see e.g. [41, 44, 94, 141]). Although the process of vascularization is strongly distributed (see e.g. [27, 45, 58]) some simple two or three compartmental models have been also proposed [40, 102]. The advantage of using antiangiogenic therapy is in resistancy to drug resistance [62]. It is due to the fact that it is directed against non-malignant endothelial cells which are genetically stable. Methodology described in our paper can be efficiently extended for this class of nonlinear models. Moreover decomposition for finite dimensional controlled part and infinite dimensional uncontrolled part used by us in analysis and optimization of drug resistance evolution and therapy may be applied to the more complicated models of angiogenesis with distributed parameter compartments.

Acknowledgements

The manuscript of this chapter was prepared when the authors were visiting the Mathematical Biosciences Institute, Ohio State University, Columbus Ohio in the fall 2003. It is our nice duty to thank Director of MBI, Profesor Avner Friedman for invitation and creating the research atmosphere. The brief version of the paper was presented at the MBI Workshop 2 on Mathematical Models of Cell Proliferation and Cancer Chemotherapy in November 2003.

We acknowledge financial support to A.S. from the International Addendum to the NSF grant DMS 0205093, from the Polish Ministry of Science grant BK 234/Raul/05 at the SUT, and from the European Commission 6th Framework Program MRNT-CT-2004-503661.

References

1. Abramowitz M., I.A. Stegun (1964) *Handbook of Mathematical Functions*, National Bureau of Standards, Washington.
2. Agur Z. (1988) The effect of drug schedule on responsiveness to chemotherapy. *Annals N.Y. Acad. Sci.* **504**: 274–277.
3. Agur Z., R. Arnon, B. Schachter (1988) Reduction of cytotoxicity to normal tissues by new regimens of phase-specific drugs. *Math. Biosci.* **92**: 1–15.
4. Alison M.R., C.E. Sarraf (1997) *Understanding Cancer-From Basic Science to Clinical Practice*, Cambridge Univ. Press.
5. Andreef M., A. Tafuri, P. Bettelheim, P. Valent, E. Estey, R. Lemoli, A. Goodacre, B. Clarkson, F. Mandelli, A. Deisseroth (1992) Cytokinetic resistance in acute leukemia: recombinant human granulocyte colony-stimulating factor, granulocyte macrophage colony- stimulating factor, interleukin-3 and stem cell factor effects in vitro and clinical trials with granulocyte macrophage colony-stimulating factor. In: *Haematology and Blood Transfusion – Acute Leukemias – Pharmacokinetics* (ed. Hidemann et al.) **34**, Springer-Verlag, Berlin, 108–116.
6. Athreya K.B., P.E. Ney (1972) *Branching Processes*. Springer, New York.
7. Axelrod D.E, K.A. Baggerly, M. Kimmel (1993) Gene amplification by unequal chromatid exchange: Probabilistic modeling and analysis of drug resistance data. *J. Theor. Biol.* **168**:151–159.
8. Bahrami K., M. Kim (1975) Optimal control of multiplicative control systems arising from cancer therapy. *IEEE Trans. Autom. Contr.* **AC 20**: 537–542.
9. Baserga R. (1985) *The Biology of Cell Reproduction*. Harvard University Press, Cambridge, MA.
10. Bate R.R. (1969) The optimal control of systems with transport lag. In: *Advances in Control and Dynamic Systems* (ed. Leondes) **7**, Academic Press, 165–224.
11. Begg A.C. (1995) The clinical status of Tpot as a predictor? Or why no tempest in the Tpot! *Int. J. Radiot. Oncol. Biol. Phys.* **32**: 1539–1541
12. Begg A.C. (2002) Critical appraisal of in situ cell kinetic measurements as response predictors in human tumors, *Semin. Radiot. Oncol.* **3**: 144–151
13. Begg A.C., N.J. McNally, D.C. Shrieve, H.A. Kärcher (1985) A method to measure duration of DNA synthesis and the potential doubling time from a single sample. *Cytometry*, **6**: 620–626.
14. Begg A.C., L. Moonen, I. Hofland, M. Dessing, H. Bartelink (1988) Human tumour cell kinetics using a monoclonal antibody against iododeoxyuridine Intratumoral sampling variations. *Radiother. Oncol.* **11**: 337–347.
15. Begg A.C., K. Haustermans, A.A. Hart, S. Dische, M. Saunders, B. Zackrisson, H. Gustaffson, P. Coucke, N. Paschoud, M. Hoyer, J.Overgaard, P. Antognoni, A. Rochetti, J. Bourhis, H. Bartelink, J.C. Horiot, R. Corvo, W. Giaretti, H.

Awwad, T. Shouman, T. Jouffroy, Z. Maciorowski, W. Dobrowsky, H. Struik-mans, G.D. Wilson (1999) The value of pretreatment cell kinetic parameters as predictors for radiotherapy outcome in head and neck cancer: a multicenter analysis. *Radiother. Oncol.* **50**: 13–23.

16. Bellman S. (1983) *Mathematical Methods in Medicine.* World Scientific, Singapore.

17. Bertuzzi A., A. d'Onfrio, A. Fasano, A. Gandolfi (2003) Regresion and regrowth of tumour cords following single-dose anticancer treatment. *Bull. Math. Biol.* **65**, 903–931.

18. Bertuzzi A., M. Faretta, A. Gandolfi, C. Sinisgali, G. Starace, G. Valoti, P. Ubezio (2002) Kinetic heterogenity of an experimental tumour revealed by BrdUrd incorporation and mathematical modelling *Bull. Math. Biol.* **64**, 355–384.

19. Bertuzzi A., A. Fasano, A. Gandolfi, D. Marangi (2002) Cell kinetics in tumour cords studied by a model with variable cell cycle lenght. *Math. Biosci.* **177-178**: 103–125.

20. Bertuzzi A., A. Gandolfi, R. Vitelli (1986) A regularization procedure for estimating cell kinetic parameters from flow-cytometric data. *Math. Biosci.* **82**: 63–85.

21. Bonadonna G., M. Zambetti, P. Valagussa (1995) Sequential of alternating Doxorubicin and CMF regimens in breast cancer with more then 3 positive nodes. Ten years results, *JAMA*, **273**: 542–547.

22. Brown B.W., J.R. Thompson (1975) A rationale for synchrony strategies in chemotherapy. In: *Epidemiology* (eds. Ludwig, Cooke), SIAM Publ., Philadelpia, 31–48.

23. Brown P.C., S.M. Beverly, R.T. Schimke (1981) Relationship of amplified Dihydrofolate Reductase genes to double minute chromosomes in unstably resistant mouse fibroblasts cell lines. *Mol. Cell. Biol.* **1**: 1077–1083.

24. de Bruijn N.G. (1958) *Asymptotic Methods in Analysis* North-Holland, Amsterdam.

25. Calabresi P., P.S. Schein (1993) *Medical Oncology, Basic Principles and Clinical Management of Cancer*, Mc Graw-Hill, New Yok,

26. Chabner B.A., D.L. Longo (1996) *Cancer Chemotherapy and Biotherapy*, Lippencott-Raven, Philadelfia

27. Chaplain M.A.J., M.E. Orme (1998) Mathematical modeling of tumor-induced angiogenesis. In: *Vascular Morphogenesis In Vivo, In Vitro, In Mente* (eds. C. Little, E.H. Sage, V. Mirinov), Birkhauser, Boston, 205–240.

28. Clare S.E., F. Nahlis, J.C. Panetta (2000) Molecular biology of breast cancer metastasis. The use of mathematical models to determine relapse and to predict response to chemotherapy in breast cancer, *Breast Cancer Res.*, **2**: 396–399

29. Cojocaru L., Z. Agur (1992) A theoretical analysis of interval drug design for cell-cycle-phase-specific drugs, *Math. Biosci.* **109**: 85–97.

30. Coldman A.J., J.H. Goldie (1983) A model for the resistance of tumor cells to cancer chemotherapeutic agents. *Math. Biosci.* **65**: 291.

31. Collins M.J., R.L. Dedrich (1982) Pharmacokinematics of anticancer drugs. In: *Pharmacologic Principles of Cancer Treatment* (ed. Chabner), Saunders, Philadelphia, 77–99.

32. Coly L.P., D.W. van Bekkum and A. Hagenbeek (1984) Enhanced tumor load reduction after chemotherapy induced recruitment and synchronization

in a slowly growing rat leukemia model (BNML) for human acute myelonic leukemia, *Leukemia Res.*, **8**: 953–963

33. Darzynkiewicz Z., F. Traganos, M. Kimmel (1987) Assay of cell cycle kinetics by multivariate flow cytometry using the principle of stathmokinesis. In: *Techniques in Cell Cycle Analysis*, (eds. Gray, Darzynkiewicz) Humana Press, Clifton, NJ, 291–336.

34. Dibrov B.F., A.M. Zhabotinsky, A. Neyfakh, H.P. Orlova, L.I. Churikova (1983) Optimal scheduling for cell synchronization by cycle-phase-specific blockers. *Math. Biosci.* **66**: 167–185.

35. Dibrov B.F., A.M. Zhabotinsky, I.A. Neyfakh, H.P. Orlova, L.I. Churikova (1985) Mathematical model of cancer chemotherapy. Periodic schedules of phase-specific cytotoxic-agent administration increasing the selectivity of therapy. *Math. Biosci.*, **73**: 1–31.

36. Doetsch G. (1964) *Introduction to the Theory and Application of the Laplace Transform*, Springer, Berlin.

37. Duda Z. (1994) Evaluation of some optimal chemotherapy protocols by using a gradient method, *Appl. Math. and Comp. Sci.*, **4**: 257–262

38. Duda Z. (1997) Numerical solutions to bilinear models arising in cancer chemotherapy, *Nonlinear World*, **4**: 53–72

39. Eisen M. (1979) *Mathematical Models in Cell Biology and Cancer Chemotherapy*, Lecture Notes in Biomathematics, Vol. 30, Springer Verlag, New York

40. Ergun A., K. Camphausen, L.M. Wein (2003) Optimal scheduling in radiotherapy and angiogenic inhibitors, *Bull. Math. Biol.* **65**: 407–424.

41. Feldman A.L., S.K. Libutti (2000) Progress in antiangiogenic gene therapy of cancer, *Cancer* **89**, 1181–1194

42. Fister K.R., J.C. Panetta (2000) Optimal control applied to cell-cycle-specific cancer chemotherapy, *SIAM J. Appl. Math.*, **60**: 1059–1072

43. von Foerster J. (1959) Some remarks on changing populations. In: *Kinetics of Cell Proliferation* (ed. Stohlman F.), Greene & Stratton, New York, 382–407.

44. Folkman J. (1975) Tumor angiogenesis, *Cancer* **3**: 355–388

45. Friedman A., F. Rietich (1999) Analysis of a mathematical model for the growth of tumors *J. Math. Biol.* **38**: 262–284

46. Goldie J.H., A.J. Coldman (1978) A mathematical model for relating the drug sensitivity of tumors to their spontaneous mutation rate. *Cancer Treat. Rep.* **63**: 1727–1733.

47. Goldie J.H., A. Coldman (1998) *Drug Resistance in Cancer*, Cambridge Univ. Press

48. Gompertz B. (1825) On nature of the function expressive of the law of human mortality, and a new mode of determining the value of life contingencies, Letter to F. Batly, Esq. *Phil. Trans. Roy. Soc.* **115**: 513–585.

49. Gray J.W. (1976) Cell cycle analysis of perturbed cell populations. Computer simulation of sequential DNA distributions. *Cell Tissue Kinet.* **9**: 499–516.

50. Gyllenberg M., G.F. Webb (1989) Quiescence as an explanation of Gompertzian tumor growth. *Develop. Aging* **53**: 25–33.

51. Hahn G.M. (1966) State vector description of the proliferation of mammalian cells in tissue culture, *Biophys. J.* **6**: 275–290.

52. Harnevo L.E., Z. Agur (1991) The dynamics of gene amplification described as a multitype compartmental model and as a branching process. *Math. Biosci.* **103**: 115–138.

53. Harnevo L.E., Z. Agur (1992) Drug resistance as a dynamic process in a model for multistep gene amplification under various levels of selection stringency. *Cancer Chemother. Pharmacol.* **30**: 469–476.

54. Harnevo L.E., Z. Agur (1993) Use of mathematical models for understanding the dynamics of gene amplification. *Mutat. Res.* **292**: 17–24.0–873.

55. Haustermans K.M., I. Hofland, H. Van Poppel, R. Oyen, W. Van de Voorde, A.C. Begg, J.F. Fowler (1997) Cell kinetic measurements in prostate cancer. *Int. J. Radiot. Oncol. Biol. Phys.* **37**: 1067–1070

56. Haustermans K.M., I. Hofland, G. Pottie, M. Ramaekers, Begg A.C. (1995) Can measurements of potential doubling time (Tpot) be compared between laboratories? A quality control study. *Cytometry,* **19**: 154–163.

57. Holmgren L., M.S. OReilly and J. Folkman (1995) Dormancy of micrometastases: balanced proliferation and apoptosis in the presence of angiogenesis suppression, *Nature Medicine,* **1**: 149–153

58. Jackson T.L. (2002) Vascular tumor growth and treatment: Consequences of polyclonality, competition and dynamic vascular support, *J. Math. Biol.* **44**: 201–216

59. Jansson B. (1975) Simulation of cell cycle kinetics based on a multicompartmental model. *Simulation* **25**: 99–108.

60. Kaczorek T. (1998) Weakly positive continuous-time linear systems, *Bull. Polish Acad. Sci.,* **46**: 233–245.

61. Kaufman R.J., P.C. Brown, R.T. Schimke (1981) Loss and stabilization of amplified dihydrofolate reductase genes in mouse sarcoma S-180 cell lines. *Mol. Cell. Biol.* **1**: 1084–1093.

62. Kerbel R.S. (1997) A cancer therapy resistant to drug resistance, *Nature* **390**: 335–336

63. Kim M., K. Brahrami, K.B. Woo (1974) A discrete time model for cell age, size and DNA distributions of proliferating cells, and its application to the movement of the labelled cohort. *IEEE Trans. Bio-Med. Eng.* **BME 21**: 387–398.

64. Kim M., K.B. Woo, S. Perry (1977) Quantitative approach to design antitumor drug dosage schedule via cell cycle kinetics and systems theory. *Ann. Biomed. Eng.* **5**: 12–33.

65. Kimmel M. (1980) Cellular population dynamics, *Math. Biosci.* **48**: pt. I, 211–224, pt. II, 225–239.

66. Kimmel M., D.E. Axelrod (2001) *Branching Processes in Biology.* Springer. NY.

67. Kimmel M., D.E. Axelrod (1991) Unequal cell division, growth regulation and colony size of mammalian cells: A mathematical model and analysis of experimental data. *J. of Theor. Biol.* **153**: 157–180.

68. Kimmel M., D.E. Axelrod (1990) Mathematical models of gene amplification with applications to cellular drug resistance and tumorigenicity. *Genetics* **125**: 633–644.

69. Kimmel M., D.E. Axelrod (1994) Fluctuation test for two-stage mutations: Application to gene amplification. *Mutat. Res.* **306**:45–60.

70. Kimmel M., D.E. Axelrod, G.M. Wahl (1992) A branching process model of gene amplification following chromosome breakage. *Mutat. Res.* **276**: 225–240.

71. Kimmel M., D.N. Stivers (1994) Time-continuous branching walk models of unstable gene amplification. *Bull. Math. Biol.* **56**: 337–357.

72. Kimmel M., A. Swierniak (1983) An optimal control problem related to leukemia chemotherapy, *Sci. Bull. Sil. Univ. Tech. (ZN Pol. Sl. s. Aut.)* **65**: 120–131 (in Polish)

73. Kimmel M., A. Swierniak, A. Polanski (1998) Infinite-dimensional model of evolution of drug resistance of cancer cells, *J. Math. Syst. Est. Contr.* **8**: 1–16

74. Kimmel M., F. Traganos (1986) Estimation and prediction of cell cycle specific effects of anticancer drugs. *Math. Biosci.* **80**: 187–208.

75. Konopleva M., T. Tsao, P. Ruvolo, I. Stiouf, Z. Estrov, C.E. Leysath, S. Zhao, D. Harris, S. Chang, C.E. Jackson, M. Munsell, N. Suh, G. Gribble, T. Honda, W.S. May, M.B. Sporn, M. Andreef (2002) Novel triterpenoid CDDO-Me is a potent inducer of apoptosis and differentiation in acute myelogenous leukemia, *Blood* **99**, 326–335

76. Kooi M.W., J. Stap, G.W. Barendsen (1984) Proliferation kinetics of cultured cells after irradiation with X-rays and 14 MeV neutrons studied by time-lapse cinematography. *Int. J. Radiat. Biol.* **45**: 583–592.

77. Kozusko K., P. Chen, S.G. Grant, B.W. Day, J.C. Panetta (2001) A mathematical model of in vitro cancer cell growth and treatment with the antimitotic agent curacin A, *Math. Biosci.* **170**: 1–16

78. Krener A. (1977) The high-order maximal principle and its application to singular controls, *SIAM J. Contr. Optim.* **15**: 256–293

79. Lampkin B.C., N.B. McWilliams, A.M. Mauer (1974) Manipulation of the mitotic cycle in treatment of acute myeloblastic leukemia. *Blood* **44**: 930–940.

80. Ledzewicz U., H. Schättler (2002) Optimal bang-bang controls for a 2-compartment model in cancer chemotherapy, *J. Opt. Theory Appl.* **114**: 609–637.

81. Ledzewicz U., H. Schättler (2002) Analysis of a cell-cycle specific model for cancer chemotherapy, *J. Biol. Syst.* **10**: 183–206

82. Luria S.E., M. Delbrück (1943) Mutations of bacteria from virus sensitivity to virus resistance. *Genetics* **28**: 491–511.

83. Luzzi K.J., I.C. MacDonald, E.E. Schmidt, N. Kerkvliet, V.L. Morris, A.F. Chambers, A.C. Groom (1998) Multistep nature of metastatic inefficiency: dormancy of solitary cells after successful extravasation and limited survival of early micrometastases, *Amer. J. Pathology*, **153**: 865–873

84. Lyss A.P. (1992) Enzymes and random synthetics, In: *Chemotherapy Source Book*, (ed. Perry), Williams & Wilkins, Baltimore, 403–408

85. Martin R.B. (1992) Optimal control drug scheduling of cancer chemotherapy, *Automatica*, **28**: 1113–1123

86. Martin R.B., K.L. Teo (1994) *Optimal Control of Drug Administration in Cancer Chemotherapy*, World Scientific, Singapore.

87. Mohler R.R. (1973) *Bilinear Control Processes with Applications to Engineering, Ecology and Medicine*, Academic Press, New York.

88. Morrow J. (1970) Genetic analysis of azaguanine resistance in an established mouse cell line. *Genetics* **65**: 279–287.

89. Murnane, J.P., M.J. Yezzi (1988) Association of high rate of recombination with amplification of dominant selectable gene in human cells. *Som. Cell Molec. Gen.* **14**: 273–286.

90. Murray J.M. (1990) Optimal control for a cancer chemotherapy problem with general growth and loss functions. *Math. Biosci.*, **98**: 273–287.

91. Neustadt W.L. (1967) An abstract variational theory with applications to a broad class of optimization problems. *SIAM J. Contr. Optim.* **5**: 90–137.

92. Noble J., H. Schättler (2002) Sufficient conditions for relative minima of broken extremals in optimal control theory, *J. Math. Anal. Appl.* **269**: 98–128

93. Norton L., R. Simon (1977) Tumor size, sensitivity to therapy, and design of treatment schedules. *Cancer Treat. Rep.* **61**: 1307–1317.

94. O'Reilly M.S., T. Boehm, Y. Shing, N. Fukai, G. Vasios, W. Lane, E. Flynn, J.R. Birkhead, B.R. Olsen, J. Folkman (1997) Endostatin: an endogenous inhibitor of angiogenesis and tumour growth, *Cell* **88**: 277–285

95. Pakes A.G. (1973) Conditional limit theorems for a left-continuous random walk. *J. Appl. Prob.* **10**: 39–53.

96. Panetta J.C., Y. Yanishevski, C.H. Pui, J.T. Sandlund, J. Rubnitz, G.K. Rivera, R. Ribeiro, W.E. Evans, M.V. Relling (2002) A mathematical model of in vivo methotrexate accumulation in acute lymphoblastic leukemia, *Cancer Chemother. Pharmacol.* **50**: 419–428

97. Panetta J.C., A. Wall, C.H. Pui, M.V. Relling, W.E. Evans (2002) Methotrexate intracellular disposition in acute lymphoblastic leukemia: a mathematical model of gammaglumatyl hydrolase activity, *Clinical Cancer Res.* **8**: 2423–2439

98. Polanski A., A. Swierniak, Z. Duda (1993) Multiple solutions to the TPVBP arising in optimal scheduling of cancer chemotherapy *Proc.IEEE Int. Conf. Syst., Man, Cybern.* le Touquet **4**: 5–8

99. Polanski A., M. Kimmel, A. Swierniak (1997) Qualitative analysis of the infinite dimensional model of evolution of drug resistance. In: *Advances in Mathematical Population Dynamics – Molecules, Cells and Man* (eds. Arino, Axelrod, Kimmel) World Scientific, Singapore, 595–612.

100. Pontryagin L.S., V.G. Boltyanskii, R.V. Gamkrelidze, E.F. Mishchenko (1964) *The Mathematical Theory of Optimal Processes*, MacMillan, New York,

101. Raza A., H. Preisler, B. Lampkin, N. Yousuf, C. Tucker, N. Peters, M. White, C. Kukla, P. Gartside, C. Siegrist, J. Bismayer, M. Barcos, J. Bennett, G. Browman, J. Goldberg, H. Grunwald, R. Larson, J. Vardman, K. Vogler (1991) Biological significance of cell cycle kinetics in 128 standard risk newly diagnosed patients with acute myelocytic leukaemia. *Brit. Journal of Hematology* **79**: 33–39.

102. Sachs R.K., L.R. Hlatky, P. Hahnfeldt (2001) Simple ODE models of tumor growth and anti-angiogenic or radiation treatment, *Math. Comput. Mod.* **33**: 1297–1308.

103. Shin K.G., R. Pado (1982) Design of optimal cancer chemotherapy using a continuous-time state model of cell kinetics. *Math. Biosci.* **59**: 225–248.

104. Simek K., M. Kimmel (2003) A note on estimation of dynamics of multiple gene expression based on singular value decomposition. *Math. Biosci.* **182**: 183–199.

105. Smieja J., A. Swierniak, Z. Duda (2000) Gradient method for finding optimal scheduling in infinite dimensional models of chemotherapy. *J. Theor. Medicine* **3**, 25–36.

106. Smieja J., A. Swierniak (2003), Different models of chemotherapy taking into account drug resistance stemming from gene amplification, *Int. J. Appl. Math. and Comp. Sci.*, **13**, 297–305

107. Smith K.A., M.B. Stark, P.A. Gorman, G.R. Stark (1992) Fusions near telomeres occur very early in the amplification of CAD genes in Syrian hamster cells. *Proc. Natl. Acad. Sci. USA* **89**: 5427–5431.

108. Speer J.F., V.E. Petrosky, M.W. Retsky, R.H.Wardwell (1984) A stochastic numerical model of breast cancer growth that simulates clinical data. *Cancer Res.* **44**: 124–130.

109. Stark G.R.(1993) Regulation and mechanisms of mammalian gene amplification. *Adv. Cancer Res.* **61**: 87–113.

110. Sullivan M.K., S.E. Salmon (1972) Kinetics of tumor growth and regression in IgC multiple myeloma. *J. Clin. Invest.* **10**: 1697–1708.

111. Sundareshan M.K., R.S. Fundakowski (1985) Periodic optimization of a class of bilinear systems with application to control of cell proliferation and cancer therapy. *IEEE Trans. Syst. Man. Cybern.* **SMC 15**: 102–115.

112. Sundareshan M.K., R.S. Fundakowski (1986) Stability and control of a class of compartmental systems with application to cell proliferation and cancer therapy. *IEEE Trans. Autom. Contr.* **AC-31**: 1022–1032.

113. Swan G.W. (1990) Role of optimal control in cancer chemotherapy, *Math. Biosci.* **101**: 237–284

114. Swan G.W. (1986) Cancer chemotherapy optimal control using the Verhulst-Pearl equation. *Bull Math. Biol.* **48**: 381–404.

115. Swan G.W. (1985) Optimal control applications in the chemotherapy of multiple myeloma *J. Math. Appl. Med. Biol.* **2**: 139–160.

116. Swan G.W., T.L. Vincent (1977) Optimal control analysis in the chemotherapy of IgC multiple myeloma. *Bull. Math. Biol.* **39**: 317–337

117. Swierniak A. (1994) Some control problems for simplest differential models of proliferation cycle, *Appl. Math. and Comp. Sci.* **4**; 223–232

118. Swierniak A. (1995) Cell cycle as an object of control, *J. Biol. Syst.*, **3**: 41–54.

119. Swierniak A. (1989) Optimal treatment protocols in leukemia – modeling the proliferation cycle. *Trans. IMACS on Sci. Comp.* **5**: 51–53.

120. Swierniak A., Z. Duda (1992) Some control problems related to optimal chemotherapy – singular solutions. *Appl. Math. and Comp. Sci.* **2**: 293–302.

121. Swierniak A., Z. Duda (1994) Singularity of optimal control problems arising in cancer chemotherapy. *Math. and Comp. Modeling* **19**: 255–262.

122. Swierniak A., Z. Duda, A. Polanski (1992) Strange phenomena in simulation of optimal control problems arising in cancer chemotherapy. *Proc. 8th Prague Symp. Comp. Biol., Ecol., Medicine*, 58–65.

123. Swierniak A., M. Kimmel (1984) Optimal control application to leukemia chemotherapy protocols design. *Sci. Bull. Sil. Univ. Techn (ZN Pol. Sl. s. Aut.)*, **74**: 261–277 (in Polish).

124. Swierniak A., A. Polanski, Z. Duda, M. Kimmel (1997) Phase-specific chemotherapy of cancer: Optimisation of scheduling and rationale for periodic protocols. *Biocybern. Biomed. Eng.* **16**, 13–43.

125. Swierniak A., U. Ledzewicz, H. Schaettler (2003) Optimal control for a class of compartmental models in cancer chemotherapy, *Int. J. Appl. Math. and Comp. Sci.*, **13**, 357–368.

126. Swierniak A., A. Polanski (1993) All solutions to the TPBVP arising in cancer chemotherapy. *Proc. 7 Symp. System, Modeling, Control*, Zakopane, 223–229.

127. Swierniak A., A. Polanski (1994) Irregularity of optimal control problem in scheduling of cancer chemotherapy. *Appl. Math. and Comp. Sci.*, **4**: 263–271.

128. Swierniak A., A. Polanski and M. Kimmel (1996) Optimal control problems arising in cell-cycle-specific cancer chemotherapy, *Cell Prolif*, **29**: 117–139.

129. Swierniak A., A. Polanski, M. Kimmel, A. Bobrowski, J. Smieja (1999) Qualitative analysis of controlled drug resistance model – inverse Laplace and semigroup approach, *Control and Cybernetics* **28**: 61–75.

130. Swierniak A., A. Polanski, J.Smieja, M. Kimmel, J. Rzeszowska (2002) Control theoretic approach to random branching walk models arising in molecular biology. *Proc. ACC Conf.* Anchorage, 3449–3453

131. Swierniak A., J. Smieja (2001) Cancer chemotherapy optimization under evolving drug resistance. *Nonlinear Analysis* **47**: 375–386.

132. Tafuri, A., M. Andreeff (1990) Kinetic rationale for cytokine-induced recruitment of myeloblastic leukemia followed by cycle-specific chemotherapy in vitro. *Leukemia* **4**: 826–834.

133. Takahashi M. (1966, 1968) Theoretical basis for cell cycle analysis, pt. I. *J. Theor. Biol.* **13**: 203–211, pt. II **15**: 195–209.

134. Tannock I. (1978) Cell kinetics and chemotherapy: a critical review. *Cancer Treat. Rep.* **62**: 1117–1133.

135. Tarnawski R., K. Skladowski, A. Swierniak, A. Wygoda, A. Mucha (2000) Repopulation of tumour cells during radiotherapy is doubled during treatment gaps. *J. Theor. Medicine* **2**, 297–306.

136. Tarnawski R., J. Fowler, K. Skladowski, A. Swierniak, R. Suwinski, B. Maciejewski, A. Wygoda (2002) How fast is repopulation of tumour cells during the treatment gaps. *Int. J. Radiot. Oncol. Biol. Phys.* **54**: 229–236.

137. Tlsty T., B.H. Margolin, K. Lum. (1989) Differences in the rates of gene amplification in nontumorigenic and tumorigenic cell lines as measured by Luria-Delbrück fluctuation analysis. *Proc. Natl. Acad. Sci. USA* **86**: 9441–9445.

138. Traganos F., M. Kimmel (1990) The stathmokinetic experiment: A single-parameter and multiparameter flow cytometric analysis. In: *Methods in Cell Biology, Flow Cytometry* **33** (eds. Darzynkiewicz, Crissman). Academic Press, New York, 249–270.

139. Varshaver N.B., M.I. Marshak, N.I. Shapiro (1983) The mutational origin of serum independence in Chinese hamster cells in vitro. *Int. J. Cancer* **31**: 471–475.

140. Webb G.F. (1993) Resonances in periodic chemotherapy scheduling. *Proc. World Congr. Nonlin. Anal. 1992*, Tampa, Florida

141. Westphal J.R., D.J. Ruiter, R.M. De Waal (2000) Anti-angiogenic treatment of human cancer: pitfalls and promises. *Int. J. Cancer* **15**: 87

142. Wheldon T.E. (1988) *Mathematical Models in Cancer Chemotherapy*. Medical Sci. Series, Hilger, Bristol.

143. White R.A., M.L. Meistrich (1986) A comment on "A method to measure the duration of DNA synthesis and the potential doubling time from a single sample." *Cytometry* **7**: 486–490.

144. White R.A., N.H.A. Terry, M.L. Meistrich, D.P. Calkins (1990) Improved method for computing the potential doubling time from flow cytometric data. *Cytometry* **11**: 314–317.

145. Windle B., B.W. Draper, Y. Yin, S. O'Gorman, G.M. Wahl (1991) A central role for chromosome breakage in gene amplification, deletion, formation, and amplicon integration. *Gene Dev.* **5**: 160–174.

146. Windle B., G.M. Wahl (1992) Molecular dissection of mammalian gene amplification: New mechanistic insights revealed by analysis of very early events. *Mutat. Res.* **276**: 199–224.

147. Zadeh L.A., C.A. Desoer (1963) *Linear System Theory. The State Space Approach.* McGraw-Hill, NY.

148. Zietz S., C. Nicolini (1979) Mathematical approaches to optimization of cancer chemotherapy. *Bull. Math. Biol.* **41**: 305–324.

Cancer Models
and Their Mathematical Analysis

Avner Friedman

Mathematical Biosciences Institute, The Ohio State University

1 Introduction

The role of a mathematical model is to explain a set of experiments, and to make predictions which can then be tested by further experiments. In setting up a mathematical model of a biological process, by a set of differential equations, it is very important to determine the numerical value of the parameters. For biological processes are typically valid only within a limited range of parameters.

In the last four decades, various cancer models have been developed in which the evolution of the densities of cells (abnormal, normal, or dead) and the concentrations of biochemical species are described in terms of differential equations. Some of these models use only ordinary differential equations (ODEs), ignoring the spatial effects of tumor growth. The models which take spatial effects into consideration are expressed in terms of partial differential equations (PDEs), and they also need to take into account the fact that the tumor region is changing in time; in fact, the tumor region, say $\Omega(t)$, and its boundary $\Gamma(t)$, are unknown in advance. Thus one needs to determine both the unknown "free boundary" $\Gamma(t)$ together with the solution of the PDEs in $\Omega(t)$. This type of problem is called a *free boundary problem*. The models described in this chapter are free boundary problems. The main concern is the spatial/geometric features of the free boundary. Some of the basic questions are: What is the shape of the free boundary? How does the free boundary behave as $t \to \infty$? Does the tumor volume increase or shrink as $t \to \infty$? Under what conditions does the tumor eventually become dormant?

In this chapter we present generic PDE models, that is, we do not specify the parameters. The results, which we shall describe, should nevertheless be useful when dealing with perhaps somewhat different models in which some or all of the parameters are determined by experiments. The reader will find such models in articles listed in references. For clarity of exposition, we have minimized, in the actual text, historic references to the literature, although we have included a few in the concluding section.

A. Friedman: *Cancer Models and Their Mathematical Analysis*, Lect. Notes Math. **1872**, 223–246 (2006)
www.springerlink.com

This chapter is intended for applied mathematicians and modelers who do not have more than a very basic knowledge in PDEs. We have also included, for the interested reader, several open problems, although many more come easily to mind. The reader who would like to pursue mathematical research in this direction will find this to be a widely open area of research, and a very exciting one too!

2 Introduction to Tumors

Cancers are characterized by the following properties: (1) They and their progeny reproduce at a faster rate than normal cells; and (2) they invade and continue to proliferate in regions normally occupied by other cells – a process called *metastasis*. Cancers are classified by the tissue from which they arise and by the type of cells involved. For example, *leukemia* is a cancer of white blood cells, *sarcoma* is a cancer arising in muscles and connective tissue, and *carcinoma* is a cancer originating from epithelial cells, that is, the closely packed cells which align the internal cavities of the body. In this chapter, we shall deal only with carcinomas.

Neoplasm, or *tumor*, is a growing mass of abnormal cells. As long as this mass remains clustered together and confined to the cavity, the tumor is said to be *benign*. If the tumor has emerged out of the cavity, by breaking out through the basal membrane and then proliferating into the extracellular matrix, or stroma, then the tumor has become *malignant*, and we refer to it as *cancer*. When cancer cells invade into the blood stream or the lymphatic vessels, they may then be transported into another location, thus creating a *secondary tumor*; this process is called *metastasis*. The *primary tumor* is the tumor in its initiated location. A primary tumor is usually traced to a single mutated cell, from which, over a period of time, a colony of cells is formed. A solid tumor may typically be detected only when it reaches a size of 1 cm; by then the tumor contains 10^9 cells, including normal cells.

DNA replication and repair is not a 100% accurate process. As a result, many gene mutations take place in the human body over one's lifetime. There is evidence that a single abnormal cell, which gives rise to a tumor, has risen through a number of genetic mutations, or *epigenetic* mutations; the latter means a change of gene expression as a result of blocking of gene promoters.

There are two ways by which a gene can become abnormal: (1) A stimulating gene becomes hyperactive, or upregulated; such an abnormal gene is called *oncogene*; and (2) An inhibitory gene becomes inactive, or downregulated; it is called a *tumor suppressive* gene, an example being the p53 gene which controls the initiation of the cell cycle.

In our mathematical models of tumor growth, we shall represent cells by their density $p(x,t)$ rather than by their number density $N(x,t)$ (except in the model (51)–(56); here, x is a point in the tumor region and t is the time variable. The relation between these two quantities is

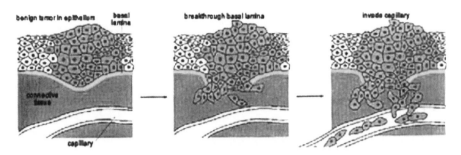

benign tumor in epithelium   basal lamina   breakthrough basal lamina   invade capillary

connective tissue

capillary

Fig. 1. The evolution of a solid benign tumor to invasive cancer

$$p(x, t) \approx N(x, t)dx$$

where $x = (x_1, x_2, x_3), dx = (dx_1, dx_2, dx_3)$, and $N(x, t)$ is the number of cells in the box with sides $(x_i, x_i + dx_i)$.

The tumor region is a 3-dimensional region $\Omega(t)$, which varies with time t. Within $\Omega(t)$ there are several types of cells, some are abnormal and others are normal, as well as several different chemicals such as oxygen and other nutrients, drugs, and immune system inhibitors. The densities of the cells and the concentrations of the chemicals satisfy a system of partial differential equations (PDEs) with appropriate boundary conditions. A major difficulty in the analysis of the models is due to the fact that the region $\Omega(t)$ is one of the unknowns of the problem; usually a physical condition is imposed on the boundary of $\Omega(t)$, which should, in principle, enable us to assert that the region $\Omega(t)$ and the solution of the system of the PDEs in $\Omega(t)$ have a unique solution. The most important component of the solution is the region $\Omega(t)$, since one would like to use the model to predict whether the tumor region will grow, whether it will invade the stroma, how long this process will take, etc.

A problem of solving a system of PDEs in a domain with an unknown boundary is called a *free boundary problem*. Such problems arise in many areas of the physical sciences and technology; for example, in solidification and melting processes where the free boundary is the solid/fluid interface, in jets emerging from a nose, and in bubbles moving in air or water. However, the free boundary problems which arise in cancer models are more complicated due to the complexity of the processes underlying the growth of tumors.

A primary tumor can grow up to a typical size of 1mm without requiring new supply of nutrients; during this stage the tumor is said to be *avascular*. In order to continue to grow, the tumor requires new sources of nutrients. It does it by secreting chemicals called *tumor growth factors*, which stimulate the formation of new blood vessels, attracting them into the tumor. This is the process of *angiogenesis*; a tumor which has developed beyond this stage is said to be *vascularized*.

Let us denote the concentration of nutrients (e.g., oxygen) in the blood by c. Then, for avascular tumors, we model the evolution of c by

$$\varepsilon_0 \frac{\partial c}{\partial t} = D_c \nabla^2 c - \lambda c \quad \text{in} \quad \Omega(t) \tag{1}$$

where $D_c, \varepsilon_0, \lambda$ are positive constants. Here λ is the nutrient consumption rate, D_c is a diffusion coefficient, and

$$\varepsilon_0 = \frac{T_{diffussion}}{T_{growth}}$$

is the ratio of the nutrient diffusion time scale to the tumor growth (e.g. tumor doubling) time scale; typically

$$T_{diffussion} \approx 1 \text{ minute}, T_{growth} \approx 1 \text{ day} ,$$

so that ε_0 is small.

For vascular tumors, we replace the equation (1) by

$$\varepsilon_0 \frac{\partial c}{\partial t} = D_c \nabla^2 c + \Gamma(c_B - c) - \lambda c \tag{2}$$

where c_B is the nutrient concentration in the vasculature and Γ is the rate of the blood-tissue transfer; thus $\Gamma(c_B - c)$ accounts for the nutrient concentration after the process of angiogenesis has taken place.

In the sequel, we shall use the change of variables, in the case of vascularized tumors,

$$c - \frac{\Gamma c_B}{\Gamma + \lambda} \to c \quad \text{and} \quad \Gamma + \lambda \to \lambda , \tag{3}$$

so that in both avascular and vascular tumors c satisfies the same equation (1). However, in the vascular case c may take negative values if $c < \Gamma c_B/(\Gamma + \lambda)$ somewhere in $\Omega(t)$ prior to the above change of variables.

3 Three Types of Cells

In this section, we assume that the tumor contains three types of cells: proliferating cells with density $p(x, t)$, quiescent cells with density $q(x, t)$, and dead cells with density $n(x, t)$. In subsequent sections we shall specialize to tumors with two types of cells, or with just proliferating cells.

A living cell becomes dead either by *apoptosis* or by *necrosis*. In apoptosis, the cell commits suicide when it receives signal to do so from the outside, or when it becomes aware of unrepairable damage to its machinery, such as its DNA. When a cell does not receive sufficient nutrients, it eventually dies, and this process is called necrosis.

Following [50], we assume that quiescent cells become proliferating at a rate $K_P(c)$ which depends on the concentration of nutrients c, and they become necrotic or go into apoptosis at death rate $K_D(c)$. We also assume that proliferating cells become quiescent at a rate $K_Q(c)$ and their death rate is

$K_A(c)$. The density of proliferating cells is increasing due to proliferation at a rate $K_B(c)$. Finally, we assume that dead cells are removed from the tumor, as they decompose, at a constant rate K_R.

Due to proliferation and removal of cells, there is a continuous motion of cells within the tumor. We shall represent this movement by a velocity field $\mathbf{v}$. We can then write the conservation of mass laws for the densities of the proliferating cells p, the quiescent cells q, and the dead cells n within the tumor region $\Omega(t)$ in the following form:

$$\frac{\partial p}{\partial t} + div(p\mathbf{v}) = [K_B(c) - K_Q(c) - K_A(c)]p + K_P(c)q , \tag{4}$$

$$\frac{\partial q}{\partial t} + div(q\mathbf{v}) = K_Q(c)p - [K_P(c) + K_D(c)]q , \tag{5}$$

$$\frac{\partial n}{\partial t} + div(n\mathbf{v}) = K_A(c)p + K_D(c)q - K_Rn . \tag{6}$$

The tumor tissue will be treated as a porous medium and the moving cells as fluid flow. In a porous medium, the velocity $\mathbf{v}$ of fluid flow is related to the fluid pressure σ by means of Darcy's law,

$$\mathbf{v} = -\beta\nabla\sigma(\beta > 0) .$$

We also assume that all the cells are physically identical in volume and mass and that their density is constant throughout the tumor. Then

$$p + q + n = const. = B .$$

For simplicity, we take $\beta = 1$ and $B = 1$. If we add equations (2)–(4), we get

$$div\mathbf{v} = K_B(c)p - K_Rn .$$

This equation can be used to replace the conservation law for n. If we also substitute $n = 1 - p - q$ in the equation for $div\mathbf{v}$, we obtain the following system of equations:

$$\varepsilon_0\frac{\partial c}{\partial t} = \Delta c - \lambda c = 0 \quad \text{in} \quad \Omega(t), t > 0 , \tag{7}$$

$$\frac{\partial p}{\partial t} - \nabla\sigma \cdot \nabla p = f(c,p,q) \quad \text{in} \quad \Omega(t), t > 0 , \tag{8}$$

$$\frac{\partial q}{\partial t} - \nabla\sigma \cdot \nabla q = g(c,p,q) \quad \text{in} \quad \Omega(t), t > 0 , \tag{9}$$

$$\Delta\sigma = -h(c,p,q) \quad \text{in} \quad \Omega(t), t > 0 , \tag{10}$$

where

$$f(c,p,q) = [K_B(c) - K_Q(c) - K_A(c)]p + K_P(c)q - h(c,p,q)p ,$$
$$g(c,p,q) = K_Q(c)p - [K_P(c) + K_D(c)]q - h(c,p,q)q ,$$
$$h(c,p,q) = -K_R + [K_B(c) + K_R]p + K_Rq .$$

We next need to impose boundary conditions. We denote the boundary of $\Omega(t)$ by $\Gamma(t)$, and take

$$c = \bar{c} \quad \text{on} \quad \Gamma(t), t > 0, \tag{11}$$

$$\sigma = \gamma\kappa \quad \text{on} \quad \Gamma(t), t > 0, \tag{12}$$

$$\frac{\partial\sigma}{\partial n} = -V_n \quad \text{on} \quad \Gamma(t), t > 0, \tag{13}$$

where $\bar{c}$ is a constant concentration of nutrients, V_n is the velocity of the free boundary, κ is the mean curvature, and γ is the surface tension coefficient. In order to explain the condition (12) we first need to explain what we mean by κ.

For any point y_0 on a surface S, consider all the curves ℓ on S passing through y_0, and denote their curvature at y_0 by $k(\ell)$. One can choose orthogonal coordinates (x_1, x_2) in the tangent plane to S at y_0, with y_0 as the origin, such that

$$k(\ell) = k_1 \cos^2\theta + k_2 \sin^2\theta$$

where θ is the angle between the x_1-axis and the tangent line to ℓ at y_0. The numbers $k_1 = k_1(y_0)$ and $k_2 = k_2(y_0)$ are called the principal curvatures, and

$$\kappa = \kappa(y_0) = \frac{1}{2}(k_1(y_0) + k_2(y_0))$$

is called the mean curvature. For a sphere S of radius R, $\kappa(y_0) \equiv R$.

The assumption (12) means that what maintains a compact solid tumor together is the surface tension. The surface tension is attributed to cell-to-cell adhesiveness [14, 15, 18].

The condition (13) means that if $x = x(t)$ describes the motion of a point on $\Gamma(t)$ and $\mathbf{n}$ is the outward normal to $\Gamma(t)$ at $x(t)$, then the normal derivative $\frac{\partial\sigma}{\partial n}$, or $(-\mathbf{v} \cdot \mathbf{n})$, is given by $-(\frac{dx}{dt}) \cdot \mathbf{n}$; this is the well known kinematic condition.

We supplement the system (7)–(13) by initial and boundary conditions,

$$p(x, 0) = p_0(x) \quad \text{and} \quad q(x, 0) = q_0(x) \quad \text{are given}$$
$$\text{functions in } \Omega(0), \quad \text{where} \quad \Omega(0) \text{ is given,}$$
$$\text{and } p_0(x) \geq 0, q_0(x) \geq 0, p_0(x) + q_0(x) \leq 1, \tag{14}$$

$$c(x, 0) = c_0(x) \quad \text{in} \quad \Omega(0), \quad c_0(x) \geq 0. \tag{15}$$

We are interested in proving that the system (7)–(15) has a "smooth" solution. If this is to be the case, then the initial data must be "smooth" and the initial and boundary data must be "consistent" with the differential equation for c at $\Gamma(0)$.

Theorem 1 [27]. *If the initial data are sufficiently smooth and the consistency conditions are satisfied, then there exists a unique smooth solution to the system (7)–(15) for some time interval $0 \leq t \leq T$.*

In general, one cannot extend the solution beyond some finite time. However if the initial data are radially symmetric, then one can prove the existence of a solution for all $t > 0$ provided $\varepsilon_0 = 0$ and the rate coefficients $K_A(c), K_B(c)$, etc. satisfy some conditions. Consider for simplicity the case where $\bar{c} = 1$, and assume that the rate coefficients are continuously differentiable for $0 \leq c \leq 1$ and satisfy the following properties:

$$K'_B(c) > 0, K'_P(c) > 0, K'_A(c) \leq 0, K'_D(c) < 0, K'_Q(c) < 0 \,,$$

$$K'_B(c) + K'_D(c) > 0 \ \ (0 \leq c \leq 1); K_B(0) = K_P(0) = 0 \,,$$

$$K_A(1) = K_D(1) = K_Q(1) = 0 \,. \tag{16}$$

The condition $K'_B(c) + K'_D(c) > 0$ is based on experimental data [29, 30], whereas all the other conditions in (16) are natural; for example, if c increases then $K_B(c)$ should increase, hence the assumption that $K'_B(c) > 0$.

Theorem 2 [28]. *If $\varepsilon_0 = 0$ and (16) holds then, for any radially symmetric initial data, there exists a unique radially symmetric solution of (7)–(15) for all $t > 0$, and the free boundary $\Gamma(t) = \{r = R(t)\}$ satisfies the inequalities*

$$\delta_0 \leq R(t) \leq A_0 \ \ for \ all \ \ t > 0 \tag{17}$$

where δ_0, A are positive constants.

Note that in the radially symmetric case $\mathbf{v} = \frac{x}{r}u(r, t)$ for some function $u(r, t)$. We also note that in the radially symmetric case we can express $R(t)$ directly by $h(c, p, q)$ by integrating (10) over $\{r < R(t)\}$ and using (13):

$$R^2(t)\frac{dR(t)}{dt} = \int_0^{R(t)} h(c, p, q)r^2 dr \,. \tag{18}$$

Since the unknown variable σ still appears in the equations (8), (9), the system for σ cannot be decoupled from the system for c, p, q.

Theorem 2 raises interesting questions:

1. Does there exist a unique radially symmetric stationary solution to the system (7)–(13)?
2. If so, does $R(t)$ converge to R_s at $t \to \infty$, where R_s is the radius of the stationary solution?
3. If there are two stationary solutions with radii R_s^- and R_s^+, which one of them is stable?
4. Are there stationary solutions which are not radially symmetric?
5. Can Theorem 2 be extended to the case where $\varepsilon_0 > 0$?

We have partial answers to these questions when the tumor contains only two types of cells, or just proliferating cells. These answers will be described in Sects. 4 and 4.

4 Two Types of Cells

In this section, we consider a special case of the model of Sect. 3, where there are only two types of cells. Suppose first that the dead cells are quickly degraded and removed from the tumor tissue. Under this assumption, we drop the equation (4) and take

$$p + q = const. = 1 .$$

In the radially symmetric case, with $\mathbf{v} = \frac{x}{r}u(r,t)$ and $\bar{c} = 1$, the system (7)–(13) with $\varepsilon_0 = 0$ takes the form

$$\frac{1}{r^2}\frac{\partial c}{\partial r}\left(r^2\frac{\partial c}{\partial r}\right) - \lambda c = 0 \quad \text{if} \quad r < R(t), t > 0 , \tag{19}$$

$$\frac{\partial p}{\partial t} + u\frac{\partial p}{\partial r} = K_P(c) + [K_M(c) - K_N(c)]p - K_M(c)p^2 \quad \text{if} \quad r < R(t), t > 0 , \tag{20}$$

$$\frac{\partial u}{\partial r} + \frac{2}{r}u = -K_D(c) + K_M(c)p \quad \text{if} \quad r < R(t), t > 0 , \tag{21}$$

$$\frac{\partial c}{\partial r}\Big|_{r=0} = 0, u|_{r=0} = 0 \quad \text{if} \quad t > 0 , \tag{22}$$

$$c = 1 \quad \text{on} \quad r = R(t) \quad \text{if} \quad t > 0 , \tag{23}$$

$$\frac{dR}{dt} = u(R(t), t) \quad \text{if} \quad t > 0 , \tag{24}$$

where $K_M = K_B - K_A + K_D, K_N = K_P + K_Q$, and we prescribe initial conditions

$$p|_{t=0} = p_0(r) \leq 1 \quad \text{if} \quad r < R(0), \ R(0) \quad \text{is given} . \tag{25}$$

From Theorem 2 we already know that the system (19)–(25) has a unique solution for all $t > 0$, and that $R(t)$ remains uniformly bounded from above and below by positive constants. So it is natural to ask whether $R(t)$ has a limit as $t \to \infty$. If this is the case, and if we set $R_s = \lim_{t\to\infty} R(t)$, then we expect $r = R_s$ to be the free boundary of a stationary solution $(c_s(r), p_s(r), u_s(r))$. This leads us to explore a more basic question, namely, does a stationary solution exist?; such a solution represents a benign, or dormant, malignancy.

Theorem 3 [27]. *There exists a unique stationary solution (c_s, p_s, u_s, R_s) of the system (19)–(24); furthermore,*

$$0 < p_s(r) < 1, p_s'(r) > 0, u_s(r) < 0, c_s'(r) > 0$$

if

$$0 < r < R_s ,$$

and

$$p_s(0) > 0, p_s(R_s) = 1, p_s'(R_s) > 0 .$$

It is known [22] that if a solution of (19)–(25) satisfies $dR/dt \geq 0$ (or ≤ 0) for all t sufficiently large, or even if just

$$\lim_{T \to \infty} \int_T^{T+1} \left| \frac{dR}{dt} \right| dt = 0 , \tag{26}$$

then the solution must converge to the stationary solution. However, this condition is not one that can easily be verified (it may not even be satisfied), so we shall ask an easier question about the asymptotic behavior of the solutions of (19)–(25):

Problem 1 (P_1). : Suppose we take initial values such that

$$c_0(r) = c_s(r) + \varepsilon c_1(r), p_0(r) = p_s(r) + \varepsilon p_1(r) \tag{27}$$

$$\text{for } r < R_s + \varepsilon R_1, \ R(0) = R_s + \varepsilon R_1 .$$

Is it true that for $|\varepsilon|$ sufficiently small the solution $(c(r,t), p(r,t), u(r,t), R(t))$ converges to the stationary solution?

But even this problem has not been solved so far. So we shall pose a yet simpler problem for which we do have a solution. If we substitute

$$c(r,t) = c_s(r) + \varepsilon c_c(r,t), p(r,t) = p_s(r) + \varepsilon p_1(r,t) ,$$
$$u(r,t) = u_s(r) + \varepsilon u_1(r,t), R(t) = R_s + \varepsilon R_1(t)$$

into the system (19)–(24), (27) and collect only the ε-order terms, we obtain a linear system for (c_1, p_1, u_1, R_1) in $\{r < R_s, t > 0\}$, which is called the *linearization of* (19)–(24), (27) *about the stationary solution.*

Theorem 4 [22]. *The linearized system has a unique global solution, and*

$$c_1 \to 0, p_1 \to 0, u_1 \to 0, R_1 \to 0$$

as $t \to \infty$.

This linear stability result may perhaps be used to solve problem (P_1).

The results of this section can probably be extended to the case where the tumor contains proliferating cells and dead cells, but not quiescent cells.

4 Proliferating Cells

In this section we assume that the tumor contains only proliferating cells. In this case $p \equiv 1$ and we are left with only two PDEs, namely (7) and (10); the latter has the form

$$\Delta \sigma = S(c)$$

where $S(c) = K_B(c) - K_A(c)$. Taking $K_B(c)$ and $K_A(c)$ to be linear functions of c, we can write

$$S(c) = \mu(c - \tilde{c})$$

where μ is a positive constant, and $\tilde{c} < \bar{c}$. To simplify the notation we scale x, t, c, and σ, and also take $\tilde{c} > 0$, so that the free boundary problem takes the form:

$$\frac{\partial c}{\partial t} - \Delta c + c = 0 \quad \text{in} \quad \Omega(t), \ t > 0 \ (\alpha > 0) , \tag{28}$$

$$\Delta \sigma = -\mu(c - \tilde{c}) \quad \text{in} \quad \Omega(t), \ t > 0 , \tag{29}$$

$$c = 1 \quad \text{on} \quad \Gamma(t), t > 0 \quad \text{and} \quad 1 > \tilde{c} > 0 , \tag{30}$$

$$\sigma = \gamma \kappa \quad \text{on} \quad \Gamma(t), \ t > 0 \tag{31}$$

$$\frac{\partial \sigma}{\partial n} = -V_n \quad \text{on} \quad \Gamma(t), \ t > 0 \tag{32}$$

with initial conditions

$$c|_{t=0} = c_0(x) \quad \text{if} \quad x \in \Omega(0), \ \Omega(0) \quad \text{is given} . \tag{33}$$

We can also make $\gamma = 1$ or $\mu = 1$ by the above scaling, but these two parameters have different biological significance:

1. $\mu(c - \tilde{c})$ may be viewed as mitotic birth rate when $c > \tilde{c}$ and death rate when $c < \tilde{c}$;
2. The surface tension coefficient γ represents the cell-to-cell adhesiveness.

As will be seen, the parameter μ/γ plays an important role in the study of the tumor boundary. For simplicity we take $\gamma = 1$.

The results of the previous sections already tell us that the system (28)–(31) has a unique solution for a small time interval (see also [7]). In the case of radially symmetric solutions, we have (cf. (18))

$$R^2(t)\frac{dR(t)}{dt} = \int_0^{R(t)} \mu(c(r,t) - \tilde{c})r^2 dr .$$

We can then solve (28), (30) together with this relation, and then proceed to solve (29), (31) for σ. In this way one establishes the existence of a unique solution $(c(r,t), \sigma(r,t), R(t))$ for all $t > 0$.

Theorem 5 [36]. *There exists a unique radially symmetric stationary solution, given by*

$$c_s(r) = \frac{R_s}{\sinh R_s}\frac{\sinh r}{r}, \qquad \sigma_s(r) = C - \mu c_s(r) + \frac{\mu}{6}\tilde{c}r^2$$

where $C = \frac{1}{R_s} + \mu - \frac{\mu \tilde{c} R_s^2}{6}$, *and* R_s *is the unique solution of the equation*

$$\tanh R_s = \frac{R_s}{1 + (\frac{\tilde{c}}{3})R_s^2} .$$

Definition 1. *Take any initial values*

$$c_0(x) = c_s(r) + \varepsilon c_1(r, \theta, \varphi) \quad \text{in} \quad \Omega(0), \ \Omega(0) : r < R_s + \varepsilon R_1(0, \vartheta)$$

where c_1, R_1 are arbitrary functions and $|\varepsilon|$ is sufficiently small, such that the smoothness and consistency conditions of Theorem 1 are satisfied. If the local solution established in Theorem 1 can then be extended to all $t > 0$, and $(c(x,t), \sigma(x,t), \Omega(t))$ converges to the radially symmetric stationary solution centered about some center x_0 as $t \to \infty$ (x_0 depends on c_1 and R_1), then we say that the stationary solution is asymptotically stable.

Since tumors grown in vitro are nearly, but not exactly, spherical, it is important to determine whether radially symmetric tumors are asymptotically stable.

Before we address this question, let us raise another one. Since in vivo one sees a variety of spatially patterned dormant malignancies, we would like to explore whether already the simplified model (28)–(32) can produce stationary solutions which are not radially symmetric. A construction of such solutions can be achieved by perturbations that produce branches of symmetry-breaking solutions:

Theorem 6 [31]. *For any integer $\ell \geq 2$, there exists a stationary solution with free boundary*

$$r = R_s + \varepsilon Y_{\ell,0}(\theta) + O(\varepsilon^2) \,,$$

and

$$\mu = \mu_\ell + \varepsilon \mu_{\ell 1} + O(\varepsilon^2)$$

for any small $|\varepsilon|$, where $Y_{\ell,0}(\theta)$ is the spherical harmonic of mode $(\ell, 0)$, namely,

$$Y_{\ell,0}(\theta) = \sqrt{\frac{2\ell+1}{4\pi}} P_\ell(\cos\theta), \ P_\ell(x) = \frac{1}{2^\ell \ell!} \frac{d^\ell}{dx^\ell}(x^2 - 1)^\ell \,,$$

and μ_ℓ is given by

$$\frac{1}{\mu_\ell} = 2R^3 \frac{I_{3/2}(R)}{I_{1/2}(R)} \frac{I_{5/2}(R)/I_{3/2}(R) - I_{\ell+3/2}(R)/I_{\ell+1/2}(R)}{\ell[\ell(\ell+1) - 2]}$$

where $R = R_s$.

Here $I_m(r)$ is the modified Bessel function

$$I_m(r) = \Sigma_{k=0}^\infty \frac{(r/2)^{m+2k}}{k! \Gamma(m+k+1)} \,.$$

We note that

$$\mu_2 < \mu_3 < \cdots, \mu_\ell \to \infty \quad \text{if} \quad \ell \to \infty \,.$$

The two-dimensional analog of Theorem 6 was proved in [37, 38]. Theorem 6 implies that the radially symmetric stationary solution is not asymptotically stable for $\mu = \mu_2 = \mu_2(R_s)$. Indeed, any of the stationary solutions established in Theorem 6 for $\ell = 2$ and $|\varepsilon|$ small lie in an arbitrarily small neighborhood of the spherical solutions if $|\varepsilon|$ is small, and they remain non-spherical for all time. On the other hand, the radially symmetric stationary solution is asymptotically stable if μ is sufficiently small [7]. But is this still true for all $\mu < \mu_2(R_s)$?

Theorem 7 [33, 34]. *There exists a function $\mu_*(R)$ such that $\mu_*(R) = \mu_2(R)$ if $R > \bar{R}$ and $\mu_*(R) < \mu_2(R)$ in $R < \bar{R}$, and such that the following holds: If $\mu < \mu_*(R_s)$ then the spherical stationary solution is asymptotically stable, and if $\mu > \mu_*(R_s)$ then the spherical stationary solution is linearly unstable.*

$\bar{R}$ is approximately 0.62207.

The bifurcation branch of stationary solutions established in Theorem 7 for $\ell = 2$ has two parts, corresponding to $\varepsilon > 0$ and $\varepsilon < 0$. We expect one part to consist of linearly stable solutions, and the other part to consist of linearly unstable solutions. One also expects that, in case $R_s < \bar{R}$, there exists a branch of periodic solutions analogous to the Hopf bifurcation. Indeed, the following holds:

Theorem 8 [35]. *(i) If $\mu_*(R) = \mu_2(R)$ then the bifurcation branch asserted in Theorem 2 for $\ell = 2$ is linearly stable for $\varepsilon > 0$, and linearly unstable for $\varepsilon < 0$; (ii) If $\mu_*(R) < \mu_2(R)$ then there is a linearly stable Hopf bifurcation at $\mu = \mu_*(R)$.*

The assertion in case (ii) means that we have a family of time-periodic solutions, and every solution of the linearized problem converges to one of these solutions as $t \to \infty$.

If $\gamma \neq 1$ in (31) then Theorems 5–8 remain valid with μ and μ_ℓ replaced by μ/γ and μ_ℓ/γ. If we think of μ as a fixed parameter and set

$$\gamma_*(R) = \frac{\mu}{\mu_*(R)} \, ,$$

then Theorem 7 asserts that the stationary spherical solution is asymptotically stable if the cell-to-cell adhesion is sufficiently strong, namely, if $\gamma > \gamma_*(R_s)$ and it is linearly unstable if $\gamma < \gamma_*(R_s)$. Theorem 6 asserts that non-spherical dormant malignancies can occur when cell-to-cell adhesion is weak, and Theorem 8 addresses the linear stability of the first branch of such non-spherical tumors.

5 Tumors with Necrotic Core

Theorem 3 shows that in a radially symmetric dormant tumor that contains only proliferating and quiescent cells, the density of the proliferating cells increases toward the boundary of the tumor. A similar result can be established for a tumor that contains only proliferating and dead cells. It has also been experimentally observed that proliferating cells are found mostly near the boundary of solid tumors, whereas necrotic cells occupy the interior of tumors. This suggests developing a mathematical model for stationary tumors in which the tumor tissue is divided into two regions: A core region Ω_0 consisting only of dead cells (the *necrotic core*), and a shell-like region Ω consisting only of proliferating cells. The boundary Γ_0 between Ω_0 and Ω is a free boundary (the *inner free boundary*).

For the time-dependent model, we introduce the notation

$$\Omega_0(t) = \text{necrotic core,}$$
$$\Omega(t) = \text{proliferating shell-like region,}$$
$$\Gamma_0(t) = \text{boundary of } \Omega_0(t),$$
$$\Gamma(t) = \text{outer boundary of } \Omega(t).$$

We assume that

$$c(x, t) \equiv const. = c_0^* \quad \text{in} \quad \Omega_0(t)$$

and that $c(x, t)$ is continuously differentiable across $\Gamma_0(t)$. Then

$$c = c_0^*, \frac{\partial c}{\partial n} = 0 \quad \text{on} \quad \Gamma_0(t) . \tag{34}$$

We also assume that the drop in the pressure σ across $\Gamma_0(t)$ is given by

$$\int_{\Gamma_0(t)} \frac{\partial \sigma}{\partial n} ds = -\nu |\Omega_0(t)| \quad (\nu > 0) \tag{35}$$

where $|\Omega_0(t)|$ denotes the volume of $\Omega_0(t)$ and $\partial/\partial n$ is the derivative in the direction of the outward normal.

In the outer shell, $\Omega(t)$, which contains only proliferating cells, we still have the system (28)–(32) with the initial condition (33), but we assume that $c_0(x)$ satisfies:

$$c_0 \Big|_{\Gamma_0(0)} = c_0^*, \frac{\partial c_0}{\partial n}\Big|_{\Gamma_0(0)} = 0, c_0^* < c_0(x) < 1 \quad \text{in} \quad \Omega(0) . \tag{36}$$

Finally, it is natural to assume that

$$c_0^* < \tilde{c} < 1 . \tag{37}$$

Problem 1 (P_2). Extend all the results of Sect. 4 to the system (28)–(33), (34)–(37).

The mathematical novelty here is that we have two free boundaries, and that the free boundary conditions on $\Gamma_0(t)$ are of a different type than the free boundary conditions on $\Gamma(t)$.

Integrating (29) over $\Omega(t)$ and using (33), (35), we obtain the relation

$$\int_{\Omega(t)} \mu(c - \tilde{c})dx = \int_{\Gamma(t)} V_n ds + \nu|\Omega_0(t)| . \tag{38}$$

We shall consider here only the case of radially symmetric solutions. We can then restate the problem in terms of only the function c. Indeed, if we set

$$\Gamma_0(t) = \{r = \rho(t)\}, \Gamma(t) = \{r = R(t)\}$$

then $c = c(r,t)$ satisfies:

$$\alpha \frac{\partial c}{\partial t} - \frac{1}{r^2} \frac{\partial}{\partial r} \left(r^2 \frac{\partial c}{\partial r} \right) + c = 0 \quad \text{if} \quad \rho(t) < r < R(t), \ t > 0 \tag{39}$$

$$c(\rho(t), t) = c_0^*, \frac{\partial c}{\partial r}(\rho(t), t) = 0, \ t > 0 , \tag{40}$$

$$c(R(t), t) = 1, \ t > 0 , \tag{41}$$

and, by (38),

$$R^2(t) \frac{dR(t)}{dt} = \int_{\rho(t)}^{R(t)} \mu(c(r,t) - \tilde{c}) r^2 dr - \frac{\nu}{3} \rho^3(t) , \tag{42}$$

with initial conditions

$$c|_{t=0} = c_0(r) \quad \text{for} \quad \rho(0) < r < R(0) \quad \text{where} \quad \rho(0), R(0) . \tag{43}$$

After we find c, we can proceed to solve for σ.
    Set

$$\gamma = \frac{\tilde{c}}{c_0^*}, \quad \gamma_0 = \frac{1}{c_0^*} \ (1 < \gamma < \gamma_0)$$

and introduce the function

$$m(\eta) = \frac{\eta \cosh \eta - \sinh \eta}{\eta^3} .$$

One can show that $m'(\eta) > 0$ for all $\eta > 0$ and

$$m(0) = \frac{1}{3}, \quad m(\infty) = \infty .$$

Hence there exists a unique number η_γ such that

$$m(\eta_\gamma) = \frac{\gamma}{3} .$$

Theorem 9 [26]. *If $\gamma_0 > \cosh \eta_\gamma$ then there exists a unique radially symmetric stationary solution $(c_s(r), \rho_s, R_s)$; The solution has the form*

$$c_s(r) = \frac{c_0^*}{r}[\sinh(r - \rho_s) + \rho_s \cosh(r - \rho_s)], \qquad \rho_s < r < R_s$$

where ρ_s and R_s are determined by the conditions $c_s(R_s) = 1$ and

$$\int_{\rho_s}^{R_s} \mu(c_s(r) - \tilde{c})r^2 dr - \frac{\nu}{3}\rho_s^3 = 0 .$$

For a class of initial data which lie in a small neighborhood of the stationary solution, the system (39)–(43) has a unique solution $(c(r, t), \rho(t), R(t))$ and it converges to $(c_s(r), \rho_s, R_s)$ as $t \to \infty$, provided α (in (39)) is sufficiently small [26].

6 Cancer Therapy

We consider here the treatment of cancer by drugs (e.g., chemotherapy). More generally, we shall use the word "inhibitor" to include not only externally administered drugs, but also chemicals produced by the autoimmune system. For simplicity we shall lump together all the inhibitors into one, and denote its concentration by u. We assume that u satisfies a diffusion equation

$$\alpha_1 \frac{\partial u}{\partial t} - \Delta u - \gamma u = 0 \quad \text{in} \quad \Omega(t), \ t > 0 \tag{44}$$

where α_1, γ are positive constants; γu represents the decay rate of the inhibitor.

The models introduced in the previous sections need to be modified by taking into account the effects of the inhibitor. We illustrate this with a drug which is designed to block the process of angiogenesis in vascular tumors. The drug causes a decrease in the concentration of nutrients. Hence, in the equation (2) we add the term $-Ku$ to the right-hand side, where K depends on the effectiveness of the drug.

Problem 2 (P_3). Extend all the results of the previous sections to the case where $-Ku$ is added to the right-hand side of (2), and u satisfies the equation (44) with prescribed boundary and initial conditions.

We note that Theorem 1 extends to this case with minor changes in the proof. We shall now consider the radially symmetric case with tumors consisting of only proliferating cells. Then

$$\alpha \frac{\partial c}{\partial t} = \frac{1}{r^2} \frac{\partial}{\partial r}\left(r^2 \frac{\partial c}{\partial r}\right) - c - Ku \quad \text{if} \quad r < R(t), \ t > 0 , \tag{45}$$

$$\alpha_1 \frac{\partial u}{\partial t} = \frac{1}{r^2} \frac{\partial}{\partial r} \left(r^2 \frac{\partial u}{\partial r} \right) - \gamma u \quad \text{if} \quad r < R(t), \ t > 0 , \tag{46}$$

$$\frac{\partial c}{\partial r} = \frac{\partial u}{\partial r} = 0 \quad \text{at} \ r = 0, \ t > 0 , \tag{47}$$

$$c|_{R(t)} = \bar{c}, u|_{R(t)} = \bar{u}, \ t > 0 , \tag{48}$$

$$R^2(t) \frac{dR(t)}{dt} = \int_0^{R(t)} \mu(c - \tilde{c})r^2 dr, \ t > 0 , \tag{49}$$

$$c|_{t=0} = c_0(r), u|_{t=0} = u_0(r) \quad \text{for} \quad r < R(0), R(0) \ \text{is given} . \tag{50}$$

As in Sect. 5, after solving this system we can proceed to solve for the pressure σ.

Note that for vascular tumors, the c which appears in (45)–(49) is not the concentration of nutrient; indeed, it is obtained from the nutrient concentration by the change of variables (3). Hence the quantities $c, \bar{c}, \tilde{c}$ and c_0 in (45)–(49) are not necessarily positive.

Theorem 10 [25]. *If $0 < (\tilde{c}/\bar{c}) < 1$ then there exists a unique stationary solution (c_s, u_s, R_s) for all $t > 0$, and, provided α, α_1 are sufficiently small, the stationary solution is globally asymptotically stable if $\bar{c} \geq 0$ and unstable if $\bar{c} < 0$.*

By "globally asymptotically stable" we mean that for *any* initial data, the solution $(c(r,t), u(r,t), R(t))$ converges to $(c_s(r), u_s(r), R_s)$; "unstable" means that for some initial data, $R(t) \to \infty$ if $t \to \infty$.

If $(\tilde{c}/\bar{c})$ does belong to the interval $(0,1)$ then there may be no stationary solutions, one stationary solution with radius R_s, or two stationary solutions with radii R_s^-, R_s^+, where $R_s^- < R_s^+$; all three cases do occur, depending on the coefficients of the system (45)–(49). When there are two stationary solutions, the one with R_s^- is asymptotically stable provided $R(0) < R_s^+$; if $R(0) > R_s^+$ then there are initial data for which $R(t) \to \infty$ as $t \to \infty$ [4].

The proofs of the above results yield additional information regarding the treatment:

(i) $R_s^- \to 0$ and $R_s^+ \to \infty$ if $\bar{u} \to \infty$. Hence, given a tumor with initial size $R(0)$, we can increase the dose $\bar{u}$ to a level $\bar{u}_*$ such that

$$R_s^- < R(0) < R_s^+ ,$$

and then $R(t)$ will decrease to R_s^- as t increases to ∞. In other words, every tumor can be made to shrink to any given small size provided it is treated with sufficiently high dose $\bar{u}$ (neglecting side-effects):

(ii) By increasing the coefficient K (i.e., the drug effectiveness), we decrease R_s^- and increase R_s^+, so that again we can make any tumor shrink to any small size.

Our next model is concerned with gene therapy. One of the obstacles in developing efficient gene therapy to cancer is caused by the delivery process. The macromolecules used as carriers to deliver the gene therapy are too large to be transported and diffused into the nuclei of the tumor cells. A recent approach aimed at bypassing this problem involves the use of virus. The virus is engineered to be replication-competent and to selectively bind to receptors on the tumor cell surface (but not to the surface of normal healthy cells). The virus particles then proceed to proliferate within the tumor cell, eventually causing death (lysis). Thereupon the newly reproduced virus particles are released and then proceed to infect adjacent cancer cells. This process continues until all the cancer cells are destroyed.

We model this process, as in [60, 61], by introducing three types of cells: cells uninfected by the virus particles, cells infected by the virus, and dead cells. Let

p = number density of uninfected cells,
q = number density of infected cells,
n = number density of dead cells,
w = number concentration of the free virus particles, i.e., the virus residing outside the cells.

Then, by conservation of mass,

$$\frac{\partial p}{\partial t} + div(p\mathbf{v}) = \lambda p - \beta pw \quad \text{in} \quad \Omega(t),\ t > 0\,, \tag{51}$$

$$\frac{\partial q}{\partial t} + div(q\mathbf{v}) = \beta pw - \delta q \quad \text{in} \quad \Omega(t),\ t > 0\,, \tag{52}$$

$$\frac{\partial n}{\partial t} + div(n\mathbf{v}) = \delta q - \mu n \quad \text{in} \quad \Omega(t),\ t > 0\,, \tag{53}$$

where λ = proliferation rate of the uninfected cells, βpw accounts for infection of uninfected cells, δ = death rate of infected cells and μ = removal rate of dead cells. When a cell dies, virus particles are released. Because virus particles are small, they satisfy a diffusion equation, so that

$$\varepsilon_0 \frac{\partial w}{\partial t} = D_w \Delta w - \gamma w + N\delta q \tag{54}$$

where $N\delta q$ is the virus release term, and γ is the virus decay rate.

As in Sect. 3 we assume that

$$\mathbf{v} = -\nabla\sigma, \quad \sigma = pressure\,.$$

Since the virus particles are very small, we ignore their contribution to the average density of the tumor tissue, and thus take $p + q + n = const. = c$. By scaling we may take $c = 1$. Then we can replace equation (53) by

$$-\Delta\sigma = \lambda p - \mu(1 - p - q) \quad \text{in} \quad \Omega(t),\ t > 0\,. \tag{55}$$

We also assume the boundary conditions (31), (32), and

$$\frac{\partial w}{\partial n} = 0 \quad \text{on} \quad \Gamma(t), \ t > 0 .$$ (56)

Theorem 1 extends to the system (51), (52), (54)–(56) and (12)–(14) with initial condition

$$w(x,0) = w_0(x) \quad \text{in} \quad \Omega(0), \ w_0(x) \geq 0 .$$

We next consider only radially symmetric solutions and set

$$\mathbf{v} = \frac{x}{r} u(r,t), M = \frac{\beta N}{\gamma}, D_w = k_0 R^2(t) ;$$

we assume that k_0 is constant, and replace w by w/N. We then obtain the system

$$\frac{\partial p}{\partial t} + u \frac{\partial p}{\partial r} = \lambda p - M\gamma p w - p h(p,q) \quad \text{if} \quad r < R(t), \ t > 0 ,$$ (57)

$$\frac{\partial q}{\partial t} + u \frac{\partial q}{\partial r} = M\gamma p w - \delta q - q h(p,q) \quad \text{if} \quad r < R(t), \ t > 0 ,$$ (58)

$$\varepsilon_0 \frac{\partial w}{\partial t} = k_0 R^2 \frac{1}{r^2} \frac{\partial}{\partial r} \left(r^2 \frac{\partial w}{\partial r} \right) + \delta q - \gamma w \quad \text{if} \quad r < R(t), \ t > 0 ,$$ (59)

$$\frac{1}{r^2} \frac{\partial}{\partial r} (r^2 u) = -h(p,q) \quad \text{if} \quad r < R(t), \ t > 0 ,$$ (60)

where

$$h(p,q) = -\lambda p + \mu(1 - p - q) ,$$

and

$$\left. \frac{\partial w}{\partial r} \right|_{r=0} = 0, \left. \frac{\partial w}{\partial r} \right|_{r=R(t)} = 0 \quad \text{if} \quad t > 0 ,$$ (61)

$$u(0,t) = 0 \quad \text{if} \quad t > 0 ,$$ (62)

$$\frac{dR}{dt} = u(R(t),t) \quad \text{if} \quad t > 0 ,$$ (63)

with initial conditions

$$p|_{t=0} = p_0(r), q|_{t=0} = q_0(r), w|_{t=0} = w_0 \quad \text{if} \quad r < R(0)$$

$$\text{where } R(0) \text{ is given} ,$$ (64)

p_0, q_0, w_0 are nonnegative

$$\text{and continuously differentiable, and } p_0 + q_0 \leq 1 .$$ (65)

Theorem 11 [39]. *There exists a unique solution $(p(r,t),\ q(r,t),\ w(r,t),\ u(r,t),\ R(t))$ of the system (57)–(65) for all $t > 0$, and*

$$R(0)e^{-\nu t} \le R(t) < R(0)e^{\nu t} \quad \text{for all} \quad t > 0,$$

for some positive constant ν.

The system (51), (52), (54), (55) with $\mathbf{v} = -\nabla\sigma$ is similar to the system (4), (5), (7), (10), but there are some important differences. For the radially symmetric case, Theorem 11 does not assume that $\varepsilon_0 = 0$, as we did in Theorem 2; on the other hand, the bounds on $R(t)$ obtained in Theorem 11 are much weaker than the bound (17) asserted in Theorem 2.

With regard to stationary solutions, for the present system, we can immediately construct such solutions with constant cell number densities

$$p(r) \equiv p_s, q(r) \equiv q_s$$

and

$$w(r) \equiv w_s = \frac{\delta}{\gamma}q_s ,$$

$$u_s(r) = \frac{1}{3}(-\mu + (\lambda + \mu)p_s + \mu q_s)r .$$

Indeed, there are four such pairs (p_s, q_s):

$$(0,0), (1,0), \left(0, \left(1 - \frac{\delta}{\mu}\right)\right) \quad \text{provided } \delta < \mu , \tag{66}$$

and

$$\left(\frac{\lambda\mu - M\delta\mu + M\delta^2 + \mu\delta}{(M\delta - \lambda)M\delta}, \frac{(\lambda + \mu)(M\delta - \delta - \lambda)}{(M\delta - \lambda)M\delta}\right) \equiv (p_s^*, q_s^*) \tag{67}$$

provided $p_s^* \ge 0, q_s^* \ge 0$.

Notice that since $u_s(R_s) = 0$, we must have $R_s = 0$. This suggests the possibility of asymptotic stability with $\lim_{t\to\infty} R(t) = 0$. Indeed we have:

Theorem 12 [39]. *Let $R(0)$ be arbitrary positive number and assume that*

$$p_0(r) - p_s, q_0(r) - q_s, w_0(r) - \frac{\delta}{\gamma}q_s$$

are uniformly small in absolute value together with their first derivative, where $(p_s, q_s) = (0, 1 - \frac{\delta}{\mu}), \delta < \mu$, and

$$M > \frac{\mu(\lambda + \delta)}{\gamma(\mu - \delta)} .$$

Then

$$\frac{dR(t)}{dt} < 0$$

and

$$R(0)e^{-\frac{\delta}{2}t} \le R(t) \le R(0)e^{-\frac{\delta}{4}t}$$

for all $t > 0$.

A similar result holds for the stationary point (67), but not for the first two points in (66).

Problem 3 (P_4). Theorem 12 establishes complete therapy if the initial concentration of (p, q) is near the point $(0, 1 - \frac{\delta}{\mu})$. Is there a general class of initial data $(p_0(r), q_0(r))$ for which one may choose $w_0(r)$ that will shrink the tumor to zero?

For other models of drug treatment see [24, 43–45, 62].

7 Concluding Remarks

A history of mathematical models of tumor growth, which includes exhaustive literature, was recently published by Aranjo and McElwain [6]. Other good sources can be found in the volume [4] edited by Adam and Bellomo and in the journal issue [42] edited by Horn and Webb.

The tumor model (28)–(32) was first developed by Greenspan [40, 41]. It was subsequently analyzed, numerically and by asymptotic analysis, by Byrne and Chaplain [16, 17] (see also [1–3, 5, 13, 47, 48] for earlier work). They studied the radially symmetric case and extended the model to include a necrotic core as in Sect. 5, and inhibitors as in Sect. 6, but with proliferating cells only. We also mention the papers by Chaplain and collaborators [19, 21, 53] and Byrne and Matthews [20] on spatial patterns in cancer growth. The model with three types of cells, as in Sect. 3, was introduced in the radially symmetric case by Pettet, Please, et al. [49]. Models with two types of cells were also considered in [46, 49, 51, 54, 59]. There are mathematical models in the literature which are designed to describe a specific type of cancer, or specific process in the evolution of general tumors, such as angiogenesis (see Chap. 2 in this volume), tumor invasion (see Chap. 4 in this volume), and the interaction of tumor with the immune system (see Chap. 3 in this volume). We mention, in particular, the work of Bertuzzi, Fasano, et al. [9–12] on tumor chords. The mathematical methods used to prove the theorems cited in this chapter may possibly be applicable to some such models. We cite one example which deals with a brain tumor such as glioblastoma, due to Sander and Deisboeck [52]; (for another example on prostate tumor, see Jackson [44]). In this model the tumor consists of a spherical core which is surrounded by invasive spherical shell of tumor cells shedded from the core in response to chemotaxis (nutrients) and homotype attraction. Denoting the invasive shell by $\{r_0 < r < R(t)\}$, the density of tumor cells by c, the nutrient concentration by n, and the homotype concentration by h, the following system of equations holds in $\{r_0 < r < R(t)\}$:

$$\frac{\partial c}{\partial t} = \nabla \cdot (D_c \nabla c) - \nabla \cdot (c \nabla (\chi n)) - \nabla \cdot (c \eta \nabla h) ,$$

$$\varepsilon_0 \frac{\partial n}{\partial t} = D_n \Delta n - c \,,$$

$$\varepsilon_1 \frac{\partial h}{\partial t} = D_h - \mu h + \lambda c \,,$$

with appropriate boundary conditions at $r = r_0$ and at the free boundary. Although this model does not fall within the models described in previous sections, it is quite possible that the mathematical methods used to prove the theorems cited in this chapter can be extended to the present model.

We finally recall that a basic assumption in this chapter regarding the physical structure of the tumor tissue was Darcy's law $\mathbf{v} = -\nabla\sigma$ where $\mathbf{v}$ is the velocity of driven cells and σ is the pressure; thus the tissue is assumed to have the consistency of a porous medium. In some cancer models it may be more appropriate to assume that the movement of cells within the tumor follows the Stokes equation for viscous flow

$$\mu \Delta \mathbf{v} = \nabla \sigma$$

where μ is the viscosity coefficient; the birth rate of tumor cells is related to $\mathbf{v}$ by

$$\nabla \cdot \mathbf{v} = K_B(c) \quad \text{or} \quad \nabla \cdot \mathbf{v} = K_B(c) - K_R n \,.$$

The Stokes equation was used by Franks, Byrne, et al. [32] to model the growth of ductal carcinoma in situ of the breast. A mathematical challenge is to develop a rigorous mathematical theory of free boundary problems for cancer models based on Stokes equation.

Acknowledgement

This work is partially supported by the National Science Foundation under Agreement No. 0112050.

References

1. Adam, J.A. (1986). A simplified mathematical model of tumor growth. *Math. Biosci.*, **81**, 224–229.
2. Adam, J.A. (1987). A mathematical model of tumor growth II: Effects of geometry and spatial nonuniformity on stability. *Math. Biosci.*, **86**, 183–211.
3. Adam, J.A. (1996). General aspect of modeling tumor growth and immune response. In J.A. Adam & N. Bellomo (Eds.), *A Survey of Models for Tumor-Immune System Dynamics* (pp. 15–87). Boston: Birkhäuser.
4. Adam, J.A., & Bellomo, N. (1997). *A Survey of Models for Tumor-Immune System Dynamics*. Boston: Birkhäuser.
5. Adam, J.A., & Maggelakis, S.A. (1990). Diffussion regulated growth characteristics of a spherical prevascular carcinoma. *Bull. Math. Biol.*, **52**, 549–582.

244   A. Friedman

6. Aranjo, R.P., & McElwain, D.L.S. (2003). A history of the study of solid tumor growth: The contribution of mathematical modelling. *Bull. Math. Biol.*, **66**, 1039–1091.

7. Bazaliy, B., & Friedman, A. (2003). A free boundary problem for an elliptic-parabolic system: Application to a model of tumor growth. *Comm. Partial Differential Equations*, **28**, 517–560.

8. Bazaliy, B., & Friedman, A. (2003). Global existence and stability for an elliptic-parabolic free boundary problem: An application of a model of tumor growth. *Indiana Univ. Math. J.*, **52**, 1265–1304.

9. Bertuzzi, A., D'Onofrio, A., Fasano, A., & Gandolfi, A. (2003). Regression and regrowth of tumour cords following single-dose anticancer treatment. *Bull. Math. Biol.*, **65**, 903–931.

10. Bertuzzi, A., Fasano, A., & Gandolfi, A. A free boundary problem with unilateral constraints describing the evolution of a tumour cord under the influence of cell killing agents. Manuscript submitted for publication.

11. Bertuzzi, A., Fasano, A., Gandolfi, A., & Marangi, D. (2002). Cell kinetics in tumour cords studied by a model with variable cell length. *Math. Biosci.*, **177 & 178**, 103–125.

12. Bertuzzi, A., & Gandolfi, A. (2000). Cell kinetics in a tumor cord. *J. Theor. Biol.*, **204**, 587–599.

13. Britton, N.F., & Chaplain, M.A.J. (1993). A qualitative analysis of some models of tissue growth. *Math. Biosci.*, **113**, 77–89.

14. Byrne, H.M. (1997). The importance of intercellular adhesion in the development of carcinomas. *IMA J. Math. Appl. Med. Biol.*, **14**, 305–323.

15. Byrne, H.M. (1999). A weakly nonlinear analysis of a model of vascular solid tumor growth. *J. Math. Biol.*, **39**, 59–89.

16. Byrne, H.M., & Chaplain, M.A.J. (1995). Growth of nonnecrotic tumors in the presence and absence of inhibitors. *Math. Biosci.*, **130**, (130–151).

17. Byrne, H.M., & Chaplain, M.A.J. (1996). Growth of necrotic tumors in the presence and absence of inhibitors. *Math. Biosci.*, **135**, 187–216.

18. Byrne, H.M., & Chaplain, M.A.J. (1996). Modelling the role of cell-cell adhesion in the growth and development of carcinomas. *Math. Comput. Modeling*, **24**, 1–17.

19. Byrne, H.M., & Chaplain, M.A.J. (1997). Free boundary value problems associated with growth and development of multicellular spheroids. *European J. Appl. Math.*, **8**, 639–358.

20. Byrne, H.M., & Matthews, P. (2002). Asymetric growth of models of avascular solid tumours: Exploiting symmetrics. *IMA Journal of Mathematics Applied to Medicine and Biology*, **19**, 1–29.

21. Chaplain, M.A.J. (1993). The development of a spatial pattern in a model for cancer growth. In H.G. Othmer, P.K. Maini, & J.D. Murray (Eds.), *Experimental and Theoretical Adavances in Biological Pattern Formation* (pp. 45–60). Plenum Press.

22. Chen, X., Cui, S., & Friedman, A. (in press). A hyperbolic free boundary problem modeling tumor growth: Asymptotic behavior. *Trans. Amer. Math. Soc.*

23. Chen, X., & Friedman, A. (2003). A free boundary problem for elliptic-hyperbolic system: An application to tumor growth. *SIAM J. Math. Anal.*, **35**, 974–986.

24. Cui, S. (2002). Analysis of a mathematical model for the growth of tumors under the action of external inhibitors. *J. Math. Biol.*, **44**, 395–426.

25. Cui, S., & Friedman, A. (2000). Analysis of a mathematical model of the effect of inhibitors on the growth of tumors. *Math. Biosci.*, **164**, 103–137.

26. Cui, S., & Friedman, A. (2001). Analysis of a mathematical model of the growth of necrotic tumors. *J. Math. Anal. Appl.*, **255**, 636–677.

27. Cui, S., & Friedman, A. (2003). A free boundary problem for a singular system of differential equations: An application to a model of tumor growth. *Trans. Amer. Math. Soc.*, **355**, 3537–3590.

28. Cui, S., & Friedman, A. (2003). A hyperbolic free boundary problem modeling tumor growth. *Interfaces and Free Boundaries*, **5**, 159–182.

29. Dorie, M.J., Kallman, R.F., & Coyne, M.A. (1986). Effect of cytochalasin b, nocodazole and irradiation on migration and internalization of cells and microspheres in tumor cell spheroids. *Exp. Cell Res.*, **166**, 370–378.

30. Dorie, M.J., Kallman, R.F., Rapacchietta, D.F., Van Antwer, D., & Huang, Y.R. (1982). Migration and internalization of cells and polystrene microspheres in tumor cell spheroids. *Exp. Cell. Res.*, **141**, 201–209.

31. Fontelos, M.A., & Friedman, A. (2003). Symmetry-breaking bifurcations of free boundary problems in three dimensions. *Asymptotic Analysis*, **35**, 187–206.

32. Franks, S.J.H., Byrne, H.M., King, J.P., Underwood, J.C.E., & Lewis, C.E. (2003). Modeling the early growth of ductal carcinoma in situ of the breast. *J. Math. Biology*, **47**, 424–452.

33. Friedman, A., & Hu, B. (2005) in press. Bifurcation from stability to instability for a free boundary problem arising in tumor model. *Archive Rat. Mech. Anal.*

34. Friedman, A., & Hu, B. (2005) in press. Asymptotic stability for a free boundary problem arising in a tumor model. *J. Diff. Eqs.*

35. Friedman, A., & Hu, B. (2005). *Stability and instability of Liapounov-Schmidt and Hopf bifurcation for a free boundary problem arising in a tumor model.* Manuscript submitted for publication.

36. Friedman, A., & Reitich, F. (1999). Analysis of a mathematical model for growth of tumors. *J. Math. Biol.*, **38**, 262–284.

37. Friedman, A., & Reitich, F. (2000). Symmetry-breaking bifurcation of analytic solutions to free boundary problems: An application to a model of tumor growth. *Trans. Amer. Math. Soc.*, **353**, 1587–1634.

38. Friedman, A., & Reitich, F. (2001). On the existence of spatially patterned dormant malignancies in the model for the growth of non-necrotic vascular tumor. *Math. Models Methods Appl. Sci.*, **11**, 601–625.

39. Friedman, A., & Tao, Y. (2003). Analysis of a model of a virus that replicates selectively in tumor cells. *J. Math. Biol.*, **47**, 391–423.

40. Greenspan, H. (1972). Models for the growth of solid tumor by diffusion. *Studies Appl. Math.*, **51**, 317–340.

41. Greenspan, H. (1976). On the growth and stability of cell cultures and solid tumors. *J. Theoret. Biol.*, **56**, 229–242.

42. Horn, M.A., & Webb, G. (Eds.). (2004). Mathematical models in Cancer. *Discrete and Continuous Dynamical Systems, Series B*, 4.

43. Jackson, T.L. (2002). Vascular tumor growth and treatment: Consequences of polyclonality, competition and dynamic vascular support. *J. Math. Biol.*, **44**, 201–226.

44. Jackson, T.L. (2004). A mathematical model of prostate tumor growth and androgen-independent relapse. *Discrete and Continuous Dynamical Systems-Series B*, **4**, 187–201.

45. Jackson, T.L., & Byrne, H.M. (2000). A mathematical model to the study of the effects of drug resistance and vasculature on the response of solid tumor to chemotherapy. *Math. Biosci.*, **164**, 17–38.
46. Landman, K.A., & Please, C.P. (2001). Tumour dynamics and necrosis: Surface tension and stability. *IMA J. Math. Appl. Med. Biol.*, **18**, 131–158.
47. Maggelakis, S.A., & Adam, J.A. (1990). Mathematical model for prevascular growth of a spherical carcinoma. *Math. Comp. Modeling*, **13**, 23–38.
48. McElwain, D.L.S., & Morris, L.E. (1978). Apoptosis as a volume loss mechanism in mathematical models of solid tumor growth. *Math. Biosci.*, **39**, 147–157.
49. McElwain, D.L.S, & Pettet, G.J. (1993). Cell migration in multicell spheroids: Swimming against the tide. *Bull. Math. Biol.*, **55**, 655–674.
50. Pettet, G.J., Please, C.P., Tindall, M.J., & McElwain, D.L.S. (2001). The migration of cells in multicell tumor spheroids. *Bull. Math. Biol.*, **63**, 231–257.
51. Please, C.P., Pettet, G.J., & McElwain, D.L.S. (1998). A new approach to modelling the formation of necrotic regions in tumours. *Appl. Math. Lett.*, **11**, 89–94.
52. Sander, L.M., & Deisboeck, T.S. (2002). Growth patterns of microscopic brain tumors. *Physical Review E*, **66**, 051901-1 to 7.
53. Sherrat, J.A., & Chaplain, M.A.J. (2001). A new mathematical model for avascular tumor growth. *J. Math. Biol.*, **43**, 291–312.
54. Thompson, K.E., & Byrne, H.M. (1999). Modelling the internalisation of labelled cells in tumor spheroids. *Bull. Math. Biol.*, **61**, 601–623.
55. Ward, J.P., & King, J.R. (1997). Mathematical modelling of avascular tumour growth. *IMA J. Math. Appl. Med. Biol.*, **14**, 36–69.
56. Ward, J.P., & King, J.R. (1999). Mathematical modelling of avascular tumour growth, II. Modelling growth saturation. *IMA J. Math. Appl. Med. Biol.*, **16**, 171–211.
57. Ward, J.P., & King, J.R. (1999). Mathematical modelling of the effects of mitotic inhibitors on avascular tumour growth. *J. Theor. Med.*, **1**, 287–311.
58. Ward, J.P., & King, J.R. (2000). Modelling the effect of cell shedding on avascular tumour growth. *J. Theor. Med.*, **2**, 155–174.
59. Ward, J.P., & King, J.R. (2003). Mathematical modelling of drug transport in tumour multicell spheroids and monolayer cultures. *Math. Biosci*, **181**, 177–207.
60. Wein, L.M., Wu, J.T., & Kirn, D.H. (2003). Validation and analysis of a mathematical model of a replication-competent oncolytic virus for cancer treatment: Implications for virus design and delivery. *Cancer Research*, **15**(63), 1317–1324.
61. Wu, J.T., Byrne, H.M., Kirn, D.H., & Wein, L.M. (2001). Modeling and analysis of a virus that replicates selectively in tumor cells. *Bull. Math. Biol.*, **63**, 731–768.
62. Wu, J.T., Kirn, D.H., & Wein, L.M. (2004). Analysis of a three-way race between tumor growth, a replication-competent virus and an immune response. *Bull. Math. Biol*, **66**, 605–625.

4. Manuscripts should in general be submitted in English. Final manuscripts should contain at least 100 pages of mathematical text and should always include

   – a general table of contents;

   – an informative introduction, with adequate motivation and perhaps some historical remarks: it should be accessible to a reader not intimately familiar with the topic treated;

   – a global subject index: as a rule this is genuinely helpful for the reader.

   Lecture Notes volumes are, as a rule, printed digitally from the authors' files. We strongly recommend that all contributions in a volume be written in the same LaTeX version, preferably LaTeX2e. To ensure best results, authors are asked to use the LaTeX2e style files available from Springer's web-server at

   ftp://ftp.springer.de/pub/tex/latex/mathegl/mono.zip (for monographs) and
   ftp://ftp.springer.de/pub/tex/latex/mathegl/mult.zip (for summer schools/tutorials).

   Additional technical instructions, if necessary, are available on request from:

   lnm@springer-sbm.com.

5. Careful preparation of the manuscripts will help keep production time short besides ensuring satisfactory appearance of the finished book in print and online. After acceptance of the manuscript authors will be asked to prepare the final LaTeX source files (and also the corresponding dvi-, pdf- or zipped ps-file) together with the final printout made from these files. The LaTeX source files are essential for producing the full-text online version of the book. For the existing online volumes of LNM see:
   http://www.springerlink.com/openurl.asp?genre=journal&issn=0075-8434.

   The actual production of a Lecture Notes volume takes approximately 8 weeks.

6. Volume editors receive a total of 50 free copies of their volume to be shared with the authors, but no royalties. They and the authors are entitled to a discount of 33.3 % on the price of Springer books purchased for their personal use, if ordering directly from Springer.

7. Commitment to publish is made by letter of intent rather than by signing a formal contract. Springer-Verlag secures the copyright for each volume. Authors are free to reuse material contained in their LNM volumes in later publications: A brief written (or e-mail) request for formal permission is sufficient.

Addresses:

Professor J.-M. Morel, CMLA,
École Normale Supérieure de Cachan,
61 Avenue du Président Wilson, 94235 Cachan Cedex, France
E-mail: Jean-Michel.Morel@cmla.ens-cachan.fr

Professor F. Takens, Mathematisch Instituut,
Rijksuniversiteit Groningen, Postbus 800,
9700 AV Groningen, The Netherlands
E-mail: F.Takens@math.rug.nl

Professor B. Teissier, Institut Mathématique de Jussieu,
UMR 7586 du CNRS, Équipe "Géométrie et Dynamique",
175 rue du Chevaleret, 75013 Paris, France
E-mail: teissier@math.jussieu.fr

For the "Mathematical Biosciences Subseries" of LNM :
Professor P. K. Maini, Center for Mathematical Biology,
Mathematical Institute, 24-29 St Giles,
Oxford OX1 3LP, UK
E-mail : maini@maths.ox.ac.uk

Springer, Mathematics Editorial I, Tiergartenstr. 17,
69121 Heidelberg, Germany,
Tel.: +49 (6221) 487-8410
Fax: +49 (6221) 487-8355
E-mail: lnm@springer-sbm.com

Printed by Books on Demand, Germany